QK941.I3V6
PLANT COMMUNITIES OF SOUTHERN ILLIN...

Plant Communities of Southern Illinois

BY THE SAME AUTHORS:

A Flora of Southern Illinois

Plant Communities

OF SOUTHERN ILLINOIS

John W. Voigt

AND

Robert H. Mohlenbrock

Southern Illinois University Press

CARBONDALE

COPYRIGHT © 1964 BY SOUTHERN ILLINOIS UNIVERSITY PRESS
LIBRARY OF CONGRESS CATALOG CARD NUMBER 64–11168
PRINTED IN THE UNITED STATES OF AMERICA
DESIGNED BY ANDOR BRAUN

PREFACE

OVER TEN YEARS AGO, with our first botanical field trips into the Southern Illinois area, came a realization that here was an area of great scenic beauty, of unusual floristic composition, interest, and overlapping vegetational patterns. Realizing that much pleasure may be obtained from recognizing and interpreting what is being viewed, this book is addressed to those whose curiosity seeks the meaning of scenery.

The plant communities described herein are each presented as a type which is determined by the conditions of its surroundings or habitat. Thus there is swamp, marsh, ditch, spring, bluff-top, prairie, lowland and upland forest, etc. Southern Illinois is predominantly forested, and many aspects of forest vegetation herewith are presented for a well-rounded picture and for varied appeal. To many, the word forest has a connotation of trees; but the forest is more than this. It is a highly constituted entity of many parts with varied interrelationships—interdependencies, discords and struggle, but also some harmonies and balance. The forest is many things. It is a home for animals and man (through lumber), a source of food, fuel, and water and also a playground. The well-developed or older forest of southern Illinois is a moist, shaded cathedral of solitude; a sparkling waterfall or spring; a dripping rock-wall plastered with delicate frond; and a spell-binding attraction ever calling whose design is ever changing. It is a living link to man's past, for it underlies his very existence and, like other natural vegetation, is a heritage too dangerously taken for granted.

Habitat conditions are forever changing and alterations occur in the communities themselves. Nature is ever changing but tends toward a dynamic stability. Man, however, induces change, too, and his activity often leads to striking alterations in the habitat. From his occupancy of the land he has imposed cutting, burning, dredging and draining, grazing and denudation. New mixtures of plants constantly are formed and struggle through a rigorous development into new communities. However, the original cover, once destroyed, is never exactly regained in its former proportion or character.

It is hoped that this book will help to expand nature's horizons for many, both locally and elsewhere. Since the book is written for the general reader, a simplified, rather historical philosophy of vegetational development and climax is presented. It is our purpose to make the material understandable to those without specialized training and, thus, only portions of our work on the vegetation of southern Illinois, which will be of interest to such an audience, is included. Representative areas were selected for portraying a particular plant community type or variation of it; much additional data could have been included, but readability would scarcely have been served in so doing. Even so, the book should be useful to the forester, to workers in the fields of wildlife, recreation, other conservation pursuits, or for basic botanical study.

Grateful appreciation is expressed to Professor Raymond J. Pool, Dr. Margaret Kaeiser, Dr. William C. Ashby, and Professor Walter B. Welch for reading the manuscript, and to Dr. Stanley Harris for reading the section on Geology. To former students David Sanders, Marvin Rensing, Wallace Weber, Carl Bollwinkel, and Donald Drapalik, we extend our thanks for their aid in gathering statistical data. For special courtesy in pointing out and leading us to certain areas, we wish to thank Mr. John Allen and Mr. Julius Swayne. Professor Charles C. Colby of the Mississippi Valley Investigations of Southern Illinois University generously permitted the making of our maps by the cartographic laboratory of that agency, and has supported the research of the section on lowland forests. Funds for photographic work have been supplied by the Research Council of the Graduate School of Southern Illinois University.

Distribution of species in figures 7, 9, and 11 were taken from Gray's Manual, eighth edition, and several state floras. We are aware that distributions as presented are of necessity incomplete but the object has been to show species with distributions which are illustrative of floristic areas, e.g., northern, southern, Appalachian, Ozarkian, etc.

Both common and scientific names are according to the nomenclature employed by Mohlenbrock and Voigt in *A Flora of Southern Illinois*. Once scientific names have been used in the text, common names are used thereafter except as presented in tables and figures.

Photographs without credit lines were made by the first author, as was figure 4. Other line illustrations were made by Mrs. Miriam Wysong Meyer.

John W. Voigt
Robert H. Mohlenbrock

Carbondale, Illinois

CONTENTS

Preface   v
List of Illustrations   xi
List of Tables   xiii
Introduction   xvii

Part I The Setting

Location and Area   3
Climate   8
Land Antiquity   21
Vegetational History and Floristics   31

Part II Vegetation

Nature's Design   51
Classification of Vegetation   68
Methods of Studying Vegetation   75

Part III Plant Communities

The Lowland Series   85
The Upland Series   134

Literature Cited   191
Index   197

ILLUSTRATIONS

1 Southern Illinois is situated in the Central Lowland near the confluence of certain major lines of drainage   4
2 A nearly geometric progression of species coming into spring bloom at Giant City State Park and a slow arithmetical monthly decline through summer and autumn   18
3 Geological map of the Southern Illinois area   24
4 A few plants related generically to those existing when dinosaurs roamed in the late Mesozoic (Cretaceous)   27
5 Maps showing advance of the four glaciers into Illinois   30
6 Heavily forested Shawnee Hills and Horseshoe Bluff in Shawnee National Forest   32
7 Alumroot (*Heuchera parviflora*) in a moist rock-wall crevice and stonecrop (*Sedum telephioides*) on a north-facing rock ledge   34
8 Appalachian affinity in distribution of alumroot and stonecrop   35
9 Harvey's buttercup (*Ranunculus harveyi*) growing on a gentle sloping upland and flower-of-an-hour (*Talinum calycinum*) growing in thin mineral soil of a sandstone rock-ledge   34
10 Ozark affinity of harvey's buttercup and flower-of-an-hour   35
11 Bishop's cap (*Mitella diphylla*) on north-facing sandstone rock and ground pine (*Lycopodium complanatum* var. *flabelliforme*) on north-facing rock shelf   36
12 Northern affinity of bishop's cap and ground pine   37
13 Examples of swamp iris (*Iris fulva*) and spider lily (*Hymenocallis occidentalis*)   38

14	Southern distribution of swamp iris and Coastal Plain affinity of spider lily	39
15	Vegetation regions of the central-states area	40
16	Zonation of a hydrosere	54
17	Layering in a forest shown by flowering redbud, and north-south slope alternation of vegetation	64
18	Representation of species distribution in a grassland quadrat	77
19	Diagram of line interception method of sampling vegetation	78
20	Hypothetical woodland situation in use of randomly selected pairs of trees	81
21	Trees of the deep swamps	86
22	Shrubby species of the deep swamps	87
23	Swamp rose (*Rosa palustris*), buttonbush (*Cephalanthus occidentalis*), and water willow (*Decodon verticillatus*) growing in swamps	88
24	Bald cypress (*Taxodium distichum*) and cypress-knee sedge (*Carex decomposita*) growing in swamps	89
25	Tupelo gum (*Nyssa aquatica*) growing in a swamp	91
26	Adventitious roots on a young green ash and example of a swamp community of box elder and silver maple	96
27	Some plants of the bottomland forests	101
28	Communities of silver maple, elm, and ash	114
29	Members of pin-oak community	117
30	Examples of rue anemone (*Anemonella thalictroides*) and bloodroot (*Sanguinaria canadensis*)	126
31	Examples of hepatica (*Hepatica acutiloba*) and valerian (*Valeriana pauciflora*)	127
32	Plants of wooded ravines	128
33	One of numerous small springs at the base of limestone bluffs at Pine Hills and site of a former spring near Wetaug	129
34	Plants of fresh-water springs	130
35	Limestone sinks as seen from the air	135
36	Two examples of drained sink-hole ponds	136
37	Sandstone and limestone escarpments in Southern Illinois	139
38	Rock ledge, Pope County, and flower-of-an-hour (*Talinum parviflorum*) which grows on rock ledges	141
39	Plants of rock ledges	142

40	Examples of woolly-lip fern (*Cheilanthes lanosa*) and of the sedge *Cyperus aristatus*	143
41	Saxifrage (*Saxifraga forbesii*) and carolina buckthorn (*Rhamnus caroliniana*)	143
42	Four species of rosin weed found in Southern Illinois	152
43	Plants of hill prairies	156
44	Blazing star (*Liatris squarrosa*), purple prairie clover (*Petalostemum purpureum*), and example of copious mulch from a meter quadrat on a hill prairie	161
45	Small hill prairie atop bluff in Saline County	163
46	Plants of upland forests	172

Tables

1	Average temperature at Southern Illinois stations and Chicago	9
2	Mean temperature and precipitation at selected Southern Illinois stations	10
3	A geological chronology of earth's history	23
4	Geographical affinities of southern plants	41
5	Percentage of frequency and composition of tree species in a *Taxodium-Nyssa aquatica/Rosa palustris* community	90
6	Percentage of frequency and composition of tree species in a *Taxodium-Fraxinus tomentosa/Itea virginica* community	93
7	Percentage of frequency and composition of tree species in a *Populus deltoides-Salix nigra/Leersia* community	99
8	Percentage of composition, based on foliage cover, and frequency of a *Populus deltoides-Salix nigra/Leersia* understory community	100
9	Vines encountered on trees, one-mile transect in bottomland woods	104
10	Vegetation-moisture table	106
11	*Acer saccharinum-Betula nigra/Saururus* community	107
12	*Quercus alba-Carya ovata/Uniola latifolia* community	107
13	Percentage composition based on basal area of bottomland hardwood forests of lower Wabash Valley	109
14	Ranking of 20 most important species of herbaceous understory of 18 bottomland hardwood areas	110

15	Shrub and tree reproduction in nine lowland areas	111
16	Percentage of frequency and composition of tree species in an *Acer saccharinum-Populus deltoides/Aster* community	113
17	Percentage of composition, based on foliage cover, and frequency of an *Acer saccharinum-Populus deltoides/Aster* understory community	115
18	Survival and growth of submerged silver maple seedlings	116
19	Percentage of frequency and composition of tree species in a pin-oak community	116
20	Percentage of frequency and composition of tree species in a *Quercus palustris/Carex hyalinolepis* community	116
21	Percentage of composition, based on foliage cover, and frequency of a *Quercus palustris/Carex hyalinolepis* community	118
22	Percentage of frequency and composition of tree species in an *Acer negundo-Platanus/Rhus radicans* community	120
23	Percentage of frequency and composition of a box elder-sycamore community	121
24	Percentage of frequency and composition, understory of an *Acer negundo-Platanus/Rhus radicans* community	122
25	Percentage of frequency and composition of tree species in a *Quercus-Carya/Hymenocallis* community	123
26	Percentage of frequency and composition of tree species in a *Fagus-Liquidambar/Rhus* community	124
27	Percentage composition of vegetation from a railroad right-of-way prairie strip	151
28	Percentage composition of vegetation from a hill prairie near Prairie du Rocher	158
29	Percentage composition of vegetation from a hill prairie in Saline County	164
30	Percentage composition of vegetation from a hill prairie in Union County	166
31	Percentage of frequency and composition and number of lowland areas in which selected species were found	169
32	Dominants in lowland, mid-slope, and upland communities	170
33	Percentage composition of trees in ravine forest communities	170
34	Association of tree species in lowlands	171
35	Percentage of frequency and composition and number of	

	areas in which selected species were present on mid-slopes	176
36	Association of species, random pairs of trees on mid-slopes	177
37	Percentage composition of trees on selected mid-slope situations	179
38	Percentage composition of tree species in selected *Pinus echinata-Carya buckleyi/Vaccinium* communities	181
39	Percentage composition of trees in selected ridge-top tree communities	183
40	Percentage of frequency and composition and number of areas in which selected species were present on uplands	184
41	Association of species, random pairs of trees on ridge tops	185
42	Shrub and tree reproduction in selected situations	187

Introduction

When one writes of the plant life of an area he may approach it in several ways. Most fundamental of these is the discovery and listing of all forms in a particular area. Such a list of plant species constitutes the flora. The construction of a manual or key for identification, such as *A Flora of Southern Illinois*, usually follows. Another approach is that of geography or ecology in which spatial distribution is emphasized. Assuming that every form has a point of origin or center from which its migrations began, we may focus attention on these centers and subsequent migration from them. Special emphasis upon migration centers and routes of migration is the viewpoint of floristic plant geography. Forces compelling migrations are sought and the time interval during which these conditions occurred is considered.

A simple discovery of shark's teeth preserved in rock layers in Europe launched the study of paleontology in the middle of the seventeenth century. Later discoveries of different kinds of fossils in still other layers of rock led to the idea of geologic succession. The earlier idea held was that these fossilized forms differed from one another from place to place because of great catastrophes which overtook living things in some places and not in others. Sir Charles Lyell (1797–1875) was the first to prove continuity of life in prehistoric or geologic times, thus dispelling the idea of catastrophic change. Lyell found evidences that the fossiliferous depositions were accumulated gradually by natural processes taking long periods of time. The fact that changes of living organisms occurred with continuity through geological time as a result of natural process was to have great influence upon the conceptual development and explanation of organic evolution.

Charles Darwin, in his memorable *Origin of Species* in 1859, showed the dynamic nature of the environment and of life's history. Darwin's work treated the importance of environment for the role it plays in success or failure of the individual and the role it plays in manifesting changed organisms through time. This mechanism of change through natural selection is now more fully explainable by the use of genetics, a discipline whose factual content was laid down by Gregor Mendel, an Austrian monk, in 1866. The work of Darwin on the environment and the work of Mendel on heredity made the appearance of ecology, as a separate branch of biology, a certainty.

The German philosopher Haeckel is generally credited with the first use of the word ecology in 1869, when he used it to refer to the body of knowledge concerning all relations of an organism and its environment. It is generally thought that the term was not in general usage until the end of the nineteenth century. Oehser (1959) reports an earlier, rather casual, use of the word by Thoreau in 1858.

In speaking of the plant life of an area from an ecological point of view the word vegetation is used rather than flora. The taxonomist, in treating the flora, is concerned with all forms more or less equally. The ecologist in his study of vegetation may be unequally concerned with the species components. His interest in a particular place at a particular time may concern, for the most part, only a few species. These few species will be those which, because of their superior stature, greater numbers, vigor, or long life, exercise a control of an area and give an expression of unity to the vegetation. Such vegetational unity long has been recognized, and there is widespread understanding of what is meant by an oak woods, cypress swamp, or wet prairie, etc. The ecologist is concerned with discovering and describing these designs of nature around the face of the earth. He is concerned with the behavior of the leading species under various conditions. Ecological studies emphasize the relationships of organisms to their environment; they emphasize what is taking place in the dynamic milieu of nature.

Our story of vegetation is, then, the story of plant life in space and time; the story of how things are and how they got that way. In explaining them, it is necessary to employ a geographic and historical approach.

PART I The Setting

Location and Area

The region of southern Illinois covered in this treatment is bounded on the west by the Mississippi River. An arbitrary northern boundary extends along 38° 15′ N. latitude from Fults in Monroe County eastward to Grayville on the Wabash River. The lower Wabash River vegetation north as far as Terre Haute, Indiana, is included. A southeastern boundary is formed by the Lower Wabash and Ohio Rivers. The southern tip of Illinois lies at about 37° N. latitude. The north-south distance of this area is about 90 miles and the east-west distance is about 126 miles. The area is roughly triangular or wedge-shaped, includes the southern seventeen counties, and has an area slightly greater than the state of Connecticut. (Fig. 1.)

The larger part of the southern Illinois area is included in the Shawnee Hills Section of the Interior Low Plateau Province. The northern part of the area lies near the southern boundary of the Till Plain Section of the Central Lowland. The western edge is a part of the Salem Plateau Section of the Ozark Plateau Province, and the southern part is in the Coastal Plain Province (Fenneman, 1938). Reflecting the strong influence of physiographic position, vegetational regions often are correlated with physiographic maps.

Illinois is known as the prairie state. This appellation derives from the fact it was the easternmost of the states with extensive covering of grassland vegetation. As grassland, it was also referred to the central plains region where the main feature was the lack of any number of striking physiographic contrasts. Relief over the state is moderate to slight. The central location and lowland nature of the region of south-

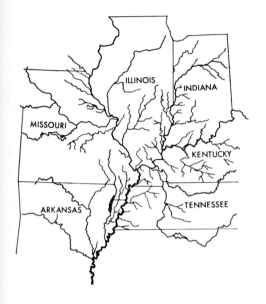

CENTRAL LOWLAND DRAINAGE PATTERN

1. Southern Illinois is situated in the Central Lowland near the confluence of certain major lines of drainage.

ern Illinois, except for the Shawneetown Ridge or escarpment, is emphasized by the confluence of the major lines of drainage (Fig. 1). Illinois is the lowest of the north-central states with a mean elevation of about 600 feet above sea level. A comparison with other surrounding states shows Indiana with 700 feet, Wisconsin 1,050 feet, Iowa 1,100 feet, and Missouri 800 feet above sea level (Gannett, 1892).

Shawnee Hills Section

The Shawnee Hills Section covers about half of Jackson County and most of Union, Johnson, Pope, and Hardin Counties. Smaller portions of neighboring counties also are included in an area of about 2,500 square miles. These Shawnee Hills stretch in an east-west direction from the western edge near the bluffs of the Mississippi River to the Ohio River on the east. The highest bedrock in the section is 1,065 feet at Williams Hill in Pope County. This location is the second highest locality in the state and the highest in southern Illinois. The bedrock floors of the large valleys are less than 200 feet above sea level, which gives a total relief of more than 850 feet (Horberg, 1950).

LOCATION AND AREA 5

Randolph County, a part of the Salem Plateau Province, lies in the northwest corner of the area treated here as southern Illinois. Several floristic affinities with neighboring areas west of the Mississippi River are known. The northern half of the county was glaciated by the Kansan advance. This land is now gently rolling. Prior to settlement it was much in prairie. There are numerous bluffs along the Mississippi River. These bluffs have loess deposits which range in thickness from ten to sixty feet. The general elevation is from three hundred to four hundred feet. Much of the area is underlain by Mississippian limestone with Pennsylvanian sandstone overlying it. Limestone is near the surface near Chester. The occurrence of numerous sink-holes in the northwest part of the county emphasizes this type of bedrock.

Southward along the Mississippi River is Jackson County, southern Illinois' largest in area. It is over 600 square miles in area and is physically one of the most diverse areas.

The terrain nearest the Mississippi River is broken or rough and hilly. A continuous line of bluffs extends along the river, or sometimes a few miles east of the river, where they overlook a broad valley which was cut and broadened by the Mississippi ages ago when other parts of this valley were once the river's course. In the southwestern part of the county an isolated, oval-shaped eminence stands in the broad valley itself. This relict land-form is called "Fountain Bluff." An erosional remnant in the old Mississippi Valley, Fountain Bluff is about three miles in length and nearly a mile-and-a-half in width (from east to west). The lower strata of this bluff are of Lower Carboniferous limestone and this is capped by a thick deposit of Caseyville sandstone. The higher portions of this area form an elevation two hundred feet above the river.

Southward over an area of bottomland and a distance of a half-mile or so occurs a tilted outcropping of Devonian strata known as "Devil's Bake Oven." Just south of this is a broken or jagged limestone outcrop, a true hogback, known as "Devil's Backbone." Nearly opposite the lower end of the "Backbone" is an outcrop of Burlington limestone which occurs on the western slope of "Walker's Hill." The upper end of Walker's Hill is composed of gray limestone which is exposed in tumbling masses on the hillsides. Fountain Bluff, the "Backbone," and Walker's Hill were all originally a part of the Missouri bluffs. It is assumed that they were isolated by erosion during Pleistocene times.

South of Grand Tower and eastward across the broad flood plain of the old Mississippi course, but along the present course of the Big Muddy River which runs southeastward to join the Mississippi, is a prominent bluff of conglomerate sandstone known as "Swallow Rock." It forms cliffs about fifty feet high at its northern end and increases in elevation southward to a perpendicular cliff about two hundred feet above the river.

Low swampy woods occur in both the western and northeastern sections of the county. Several artificial lakes (Lake Murphysboro, Campbell's Lake, etc.) provide stations for some unusual aquatics. Parts of the county formerly were covered by grasslands (Elk Prairie), but now the prairie vegetation is confined to remnants along railroads or atop the limestone bluffs (Mohlenbrock, 1959).

South of Swallow Rock near LaRue in Union County is a high, sheer bluff of Bailey limestone which reaches a thickness over three hundred feet. The Bailey limestone is an impure limestone of Devonian age. The bluff at this location is popularly referred to as the "Pine Hills" because of local growth of southern yellow pine (*Pinus echinata*) on the upper south-facing slopes of some of the hills.

The northern part of Union County is crossed by the Shawneetown Ridge, the Pennsylvanian escarpment, which forms a watershed divide extending eastward across the southern tip of Illinois to the Ohio River. "Drainage northward is by numerous short tributaries into subsequent valleys of Crab Orchard Creek and the South Fork of the Saline River. Streams south of the divide flow across the Mississippi Plain into the Cache Valley which is an abandoned channel of the Ohio River" (Horberg, 1950).

Johnson, Pope, and Hardin Counties present a continuation of the Shawneetown Ridge eastward. This ridge divides the drainage of the counties. Cache River and Bay Creek run southward, and the Big Muddy and Saline Rivers run northward.

Coastal Plain

South of the Shawneetown Ridge in southern Alexander, central and southern Pulaski, southern Massac, and Pope Counties, the land flattens and drainage is poor. There are frequent inundations and overflows. These lands may be described best as swampy. In terms of the vegetation occupying them, we would call them cypress swamps. "These swamps are surrounded by low ridges and swells. Single ponds, connected with the main body by bayous, or merely by low depressions of the ground, extend as outliers between the higher hills" (Engelmann, 1868). Before glaciation, the Cache River valley was the outlet of the Ohio River to the Mississippi. Today the Cache is a sluggish stream spread widely in times of flood. It may still, in times of flood, drain waters of the Ohio across southern Illinois into the Mississippi. It was in relatively recent time that the Ohio has come to occupy its present position (Leighton, Ekblaw, and Horberg, 1948).

Wabash Lowland

The Lower Wabash area is delimited by Parker (1936) and by Deam (1940) as a narrow strip of lowland extending from where the Wabash empties into the Ohio northward to Parke or Vigo County in Indiana. The river becomes the Illinois-Indiana boundary here, at a point below Terre Haute, and extends a tortuous 200 miles or more to join the Ohio. On the Illinois side the northward extension is to Clark County. Jones (1950, 1955) treats essentially the same area in Illinois as the Wabash Border and includes the adjacent uplands as well. The name Wabash Lowland has been applied to the tract extending on both sides of the Wabash River in the area south of the outermost Wisconsin or Shelbyville moraine. This area occupies nearly 10,000 square miles. The great levelness of the area is directly and indirectly due to glaciation and is a distinctive characteristic of the area (Logan, 1922).

A drainage area of nearly two-thirds of Indiana and a small part of adjacent Illinois contributes a heavy drainage load to the lower Wabash River (Visher, 1944). That this area has a history of flooding is well known. These are the lands known in the days of George Rogers Clark as the "drowned lands of the Wabash." Flood-marks higher than could be reached from horseback were noted on trees in summer by Thomas (1816).

Width of the floodplain and its immediate terraces varies from very narrow strips to wider expanses of several hundred yards or, in places, greater than a mile or two.

Climate

Since land heats more rapidly in summer and cools more rapidly in winter than waters of the ocean, an extreme of heat in summer and of cold in winter develops in the remote interior of continents. Lying well to the interior of the United States, Illinois displays this continental type of climate.

The climate of Illinois and particularly of southern Illinois exhibits many weather vagaries. Here at the great continental cross-roads, here in the great saucer-shaped basin of the lowland interior, is presented a continual tug-of-war between northern cold fronts and southern warm air masses. Without the protection of natural barriers of any consequence, winds are continually moving and mixing the air masses of other regions, each of which differs notably in moisture and temperature quality.

Cold air masses spawned over arctic areas have a greater density and hence a movement southward to lower elevation. Upon contact with a southern warm air mass, the heavier cold air wedges under the warm air driving the latter upward. Thus the moist, warm air is cooled at its greater heights. The beaded weight of its moisture becomes too great for the cloud matrix to bear and rain falls rather abundantly upon southern Illinois from November until late June. The storm systems of summer tend to be weaker and remain farther northward so that beginning in early July a continental air mass moves up in an easterly direction across the continent and prevents north and south air masses from coming together. A drying situation begins which is only occasionally interrupted by thundershowers of short duration. Late summer and autumn present dry days of pleasant temperature and brightness.

Illinois, a state with a long north-south axis, spans about three hundred eighty miles, or over five degrees of latitude. This great length places its northern end at a latitude equal to that of southern New Hampshire to the east and southern Oregon to the west. The southern tip of Illinois is at a latitude equal to that of Richmond, Virginia, or Oakland, California. Cairo, near the southern tip, has an average January temperature 11.1° F. higher than that of Chicago. The average July temperature reveals Chicago to be 5.8° F. cooler than Cairo. Annual temperature contrasts show Chicago at 49.8° F. and Cairo at 58.2° F. (Table 1).

1. *Average temperature (F.) at Southern Illinois stations and Chicago*

	January	July	Annual	Maximum	Minimum
Chicago	25.7	73.9	49.8	105	−23
Mt. Vernon	32.5	78.6	55.5	114	22
Carbondale	35.3	80.0	57.6	113	24
Cairo	36.8	79.7	58.2	106	−16

From "Climate of Illinois" in *Climate and Man*, 1941.

The average growing season of southern Illinois ranges between one hundred eighty-five days to two hundred days or more annually. The last killing frost in spring occurs about April 10, and the first killing frost around October 24.

In northern Illinois, soil may freeze to a depth of three feet in winter, and the ground may be snow-covered for weeks at a time. In the southern extremity of the state, snowfall is only occasional and usually lasts but a few days. Temperatures in the south fall to zero on the average of about one day each winter and soil freezes but to a depth of eight to twelve inches. Great variations prevail in the duration of soil-frost periods (Joos, 1959).

Storms

In winter and early spring the southern section of Illinois is visited by frequent rains and winds of considerable velocity. This is due apparently to low pressure systems originating in southwestern states and following a cyclonic path toward the Ohio valley. Tornado-strikes in southern Illinois are infrequent, though warnings are numerous during the four-month period from March through June each year. The storms generally pass to the south in Arkansas or northward to the prairie section of Illinois.

Precipitation

Southern Illinois, because of prevailing southern breezes and proximity to source of moisture in the Gulf of Mexico, receives the heaviest rainfall in the State. Very generally along the eastern edge of Illinois the rainfall diminishes northward at the rate of about two inches for every thirty miles.

Average annual precipitation in southern Illinois is about 44 inches, with some years ranging as low as 22 inches and others reaching as high as 74 inches. Precipitation is usually well distributed during the year with about 30 per cent in spring, 26 per cent in summer, 23 per cent in fall, and 21 per cent in winter (from *Climate and Man*, 1941). The western part of the Illinois Ozarks receives the most precipitation. The wettest month in southern Illinois varies with location (Table 2).

2. *Mean temperature (F.) and precipitation (inches) at selected Southern Illinois stations*

	Carbondale		Cairo		Harrisburg		Mt. Vernon	
January	36.0	3.87	37.4	4.48	37.0	4.06	33.9	3.44
February	38.6	2.98	40.6	3.33	39.4	3.06	36.6	2.61
March	46.9	4.40	49.0	4.72	47.5	4.55	45.0	3.98
April	57.2	4.29	60.0	3.96	57.7	4.07	56.0	4.42
May	66.3	4.31	68.8	4.09	66.9	3.86	65.7	4.27
June	75.9	4.29	77.8	3.84	76.3	3.60	75.7	4.11
July	79.3	2.83	81.2	2.60	79.8	3.13	79.5	2.91
August	77.9	3.83	79.5	3.21	78.2	3.44	77.9	3.81
September	70.5	3.60	73.0	3.32	71.0	3.06	70.6	3.56
October	59.8	3.34	62.0	2.97	60.4	2.98	59.5	3.30
November	46.1	3.48	48.6	3.80	47.0	3.14	44.9	3.39
December	37.5	2.87	39.7	3.33	38.1	2.87	35.5	2.57
Annual	57.7	44.09	59.8	43.65	58.3	41.82	56.7	42.37

Averages for period 1931–55, from Joos, 1959.

The Shawneetown Ridge acts as a barrier to southern breezes and produces convection, thus increasing the number of showers in the summer months (*Potential Water Resources of Southern Illinois*, 1957).

Climate and Plant Distribution

The influences of climate upon plant growth and distribution are well known to the extent that none of us would look for cactus or other desert plants in a wet place or conversely. Most of us are aware of some crude climate zonation and accompanying vegetational type northward from the equator as recognized long ago by Humboldt. Adjustments by plants to these varied climatic conditions are vital to their continued existence in a fixed environmental situation. Thus plant distribution depends upon a plant's physiological tolerance.

Flower production lies within certain limits of day-length condition for many plants or within certain temperature limits for some. Day and night temperature variations (thermoperiodicity) may influence functional flower production (Went, 1944) and may also influence germination or some other vital function. Complex interactions of the individual climatic factors such as temperature, moisture, light and/or others, may occur to make the problem of cause and effect most difficult to analyze. This problem is made difficult by the interactions of factors in such a way that any change of one factor intensity registers changes in the influence of the other factors. Experimental growth chambers for control of the factors separately are helping to explain these interactions.

Relative to certain factors, it may be mentioned that every plant has a zero point of intensity for each of its vital functions and conversely there is a maximum point of intensity for such functions. Plant life can only be maintained between these two points of physiological tolerance.

This physiological tolerance range or ecological amplitude is a genetically seated individual characteristic of the species and does much to explain migrations and distribution of plants. If species with similar physiological limitations exist together in an area, their proportionate distribution is determined by competition between them. The edge or range boundary of a species is imposed by a limitation of some factor or factors for the vital functioning of that species. Range boundaries, then, are imposed by the extremes of a factor or factors.

Abundant statistical information is available concerning various features of gross climatic patterns, but these kinds of data when presented as averages show little correlation with plant range boundaries. The reason for this poor correlation is easily seen in an example of an average of 80° F. in which in one instance it is the average between 140° and 20° and in another instance the average between 90° and 70°. Most plants would not grow at the upper extreme or the lower extreme in the first example while the range presented in the second example would be very suitable for many plants. Thus it is the extremes and not the means which test the plants' physiological limits. This is more understandable when one reflects that seasonal or yearly averages are taken from figures relating both to when plants grow and when they don't grow. A story emphasizing the unreliability of averages for such purposes as we have been discussing is that of the statistician who was drowned while wading across a river whose average depth was two feet.

A closer correlation of species range boundaries and the limits of glaciation may be observed in southern Illinois. Some species seem to extend northward to the glacial border, while others more northerly in their distribution are found along the glacial border but with some disjunction of range to the more northerly occurrences. The reasons for such glacial border affinity by some species may be several but undoubtedly microclimatic influences are strongly manifest from the greater topographic contrasts which are presented immediately south of the glacial border.

Microclimate

Over glaciated land there is a flatness or an undulating topography which manifests less microclimatic contrast and influence than is present in the more dissected and unglaciated land. Macroclimate is the broad climatic picture drawn from weather bureau records taken in scattered places. Microclimate by contrast is the actual climate of a local area and this may contrast greatly with the former. A very great temperature differential occurs between a ravine bottom and its adjacent upland ledge. On a July day at Giant City State Park, a 25° F. difference is noted in a blufftop temperature and the shaded underside of an overhanging rock. Such a difference is surely of considerable importance in the explanation of plant range boundaries.

Ecotypic variation

Not all plants even of the same species have the same physiological tolerances. A certain amount of natural variation occurs and it may not be outwardly visible. If the variation has survival value it is perpetuated in future generations. Thus plants are adapted to their habitats or places in which they grow. Such adapted plants may not perform exactly the same in a different home; for example, some are adjusted to growth under day-length of certain latitudes, but will not flower under other day-length conditions even though other conditions may be favorable such as temperature and moisture.

A plant which has some genetically seated morphological and/or physiological adaptive variation from other individuals of the same species is called an ecotype. Ecotypes are perhaps most easily demonstrated in a wide-ranging species which extends through several latitudinal zones or climates. In such cases different photoperiods and degree of winter hardiness have been found. Wide ranging species similarly may show different ecotypes eastward and westward.

Endemics

Some species presently restricted in range may be the only remaining colonies of plants formerly more widespread but which have suffered some catastrophe to eliminate most others of their kind. These relicts are often represented by one or only a few ecotypes. A limited migration of a species beyond a geographic barrier may also produce a colony with little genetic potential for diversity.

In southern Illinois our only true endemic of vascular plants, French's shooting star (*Dodecatheon frenchii*), may be a species of the above nature. Its known habitat is cave-like sandstone rock shelters which face east, north, or south and under which the sandy soil is very moist during its growing season in spring and early summer. This species occupies a belt less than ten miles in width across the Shawneetown Ridge, and is found nowhere else in the country.

Phenology

Calendar dates with added statistics, chiefly concerning climate, make interesting but tedious reading in the old-fashioned almanac. A much more interesting story on climate is told in the almanac of plants. Records kept on blooming dates show that plants are faithful in the appearance of their blooms and usually appear with opening buds, unfolding leaves, or developing fruits closely on schedule except as noticeable variations in climate occur. When such variation occurs we may say that spring is early or spring is late. Observation has taught us that spring may be late for some species and early for others and depends upon the weather conditions of the year.

These vegetative or reproductive activities which are especially marked or recognizable are known as phenological events. Phenological observations not only are interesting to record, but they illuminate the picture of local climates. During the past two years lists have been posted on our kitchen bulletin board for acquainting two young boys with "signs of spring." Some of the entries are as follows:

Yellow crocus in bloom	Feb. 12, 1960
Yellow crocus leaves 1 inch high	Feb. 12, 1961
Yellow crocus in bloom	Feb. 20, 1961
Bloodroot in bloom	March 30, 1960
	March 11, 1961
Spring beauty in bloom	March 29, 1960
	March 11, 1961
Turtle doves singing	March 26, 1960
	Feb. 27, 1961
Geese flying north	March 25, 1960
	March 3, 1961
Cleft phlox in bloom	April 4, 1960
	March 14, 1961

With regard to these records, it may be seen that for crocus spring of 1960 was earlier than 1961, while the reverse was true for most other species during these years.

Magnolias and forsythia, species originally from China, bloomed during April. Wild geranium, phacelia, columbine, wild ginger, buttercups, and other natives were observed. Among the birds a brown thrasher was seen on April 4. Spring came with a rush in April, but retreated briefly into the "blackberry winter" of May. This designation of a cold and rainy period lasting from one to two weeks in early May comes from the noticeable blooming of various species of *Rubus*. These observable correlations of plant and animal growth or activity are phenological events and certain ones are of common observance and long standing use. Farmers, for example, have long used "oak leaves the size of squirrels' ears" for corn-planting time. Children of the prairies look for the first pasque flower as the event heralding the springtime and the biblical quote from the Book of Solomon relates the coming of spring in these words:

> *For lo the winter is past*
> *The rains are over and gone*
> *Flowers spring up over the earth*
> *The time for singing is at hand*
> *And the voice of the turtle dove is*
> *heard over the land.*

The growing season in southern Illinois cannot be said to be the frost-free season, for some plants do, indeed, grow the year around. Chickweed, dandelion, and shepherd's purse have been found on occasion to bloom during December or January. As these pages are written, the mid-January temperature is sixty degrees and crocus leaves are an inch out of the ground.

The most obvious factor of influence upon phenological events is temperature. This factor is perhaps so readily cited as an influence because of our long wait through the winter for warmer days. All are familiar with the slowness of things during cold conditions. The speed of a plant's activity as conditioned by temperature follows certain chemical and physical laws.

All living organisms must reach a certain stage of development to reproduce, and if a plant is to do any growing it must logically have suitable temperatures. This led to an early appraisal of a plant's need for a certain number of heat hours, days, or degrees as a sum total above its minimal starting point for growth. It was on this basis that Merriam (1898) first devised his life zones for distribution of plants and animals and for growth of crops. It was soon discovered that relative to distribution of native plants, moisture was just as important and conditioned the vegetative processes and to some extent flowering. Thus temperature remaining fixed, a plant may flower early or late as conditioned by soil moisture, or other interacting environmental factors.

Length of Day

The planting of a certain kind of plant at two different times (a few days or a week or two apart) and having them flower at the same time may be an experience we have had, and we may have wondered how this can happen. We may also wonder why certain house plants do not flower when they seem to be vegetatively well off with the temperature and the watering they receive from our care. These and other problems connected with flowering have an explanation in a phenomenon known as photoperiodic response or response to day-length conditions. The understanding of day-length period dates back to the work of Garner and Allard (1920). Many studies made subsequently on other plants led to the establishment of the following categories of plants: *long-day* plants, which flower under long day-length conditions; *short-day* plants, which flower under short day-length conditions; and *indeterminate* plants, which are insensitive to either long or short days and bloom under both conditions.

Many plants are adapted to very definite day-length requirements with upper and lower critical limits for flowering. For example, plants from southern United States, the southern part of a species range, may require short days, or intermediate conditions, with an upper critical day-length of about 14 hours. On the other hand, plants of the same species from northern United States, the northern part of the species range, may require longer days and will only flower if the day-lengths are greater than 14 hours. In the former instance we would classify the 13 hour plants as short-day plants and the northern ones as long-day plants.

Not long after many plants had been exposed to various day-length conditions and appropriately placed in the above categories, a night watchman innocently lighted an area in which plants were growing. He did this each night to punch his time clock. When these plants failed to perform as expected, an investigation was made of all possible causes and finally the finger of suspicion was soon on the momentary lighting of an otherwise continuous dark period. Experimentation revealed that the total length of the night period or darkness without interruption is most important to flowering. Plants might now more reasonably be called long-night and short-night plants.

All indications point to something being produced in the leaves which promotes flowering. Such a substance which is produced in one place and has an influence in another part of the plant is called a hormone. This flower-promoting substance tentatively is being called florigen.

This understanding gives a partial explanation for distribution of plants at different latitudes. Short-day plants are found more commonly toward the equator and long-day plants at the higher latitudes where summer days are longer. The indeterminates are wide-ranging species. This information can also be put to practical use. Plants can be made to flower out of season by controlling day-length, and increased production from vegetative parts may also be achieved. The energy normally used in production of flowers is thus stored in the vegetative parts.

The Seasons

Frost and often light snow are yet on the fallen leaves in February, and very few plants bloom, but amid these leaves bravely pushes a small white-flowered plant. Long recognized for heralding the spring season, this plant has come to be known as the "harbinger-of-spring" (*Erigenia bulbosa*). Eleven other species have been found with it in first recorded bloom in our area. Faithful to February, though near its end, bloom the maples, both the silver and the red. At the edges of woods where brush has developed as a consequence of tree-cutting will be found hazelnut whose branches are adorned with dangling catkins. Tiny female flowers consisting of little more than reddish-purple, eighth-inch long, forking stigmas are found at the tips of branches. No such nearness to floral awakening is evident in northern Illinois at this date. It will be some three weeks for this story to be told there and the number of species most probably will be fewer or perhaps different from the twelve which bloomed in southern Illinois. The advance or retardation of the season may be calculated by allowing, according to Hopkins' bioclimatic law (1938), four days for each degree of latitude (approximately sixty-nine miles), and five degrees of longitude.

The winds of March are busily sweeping away the vestiges of winter's gray and tattered veil. Intermittently openings of the blue vault above the clouds show through. Each day the sun now climbs higher and the blue sky widens. Skies are now and then recovered by the uncertain weather during this month of variable mood. Through all of its flirtation with spring, through all of its advances and hesitations, the weather of March does command the appearance of wake robin, spring beauty, pepper-root, dutchman's breeches, squirrel corn, bloodroot, and other woodland herbs. A cautious number of plants totalling thirty-four have been counted at Giant City State Park during this predominantly pre-vernal month (Mohlenbrock, 1953).

March has swept the sky, but April washes it down. The soil everywhere is charged to capacity with moisture. Under the warming leaf mold, life is stirring in every underground bud, bulb, and rhizome. In the warm sunlit areas at forest edges and along gravelly bluff-tops or sandy roadsides appear tongues and patches of cleft phlox, carpets of mist flower, and spatterings of purple delphinium. Layers of redbud and dogwood may be clearly seen at the roadside under the branches of taller trees whose leaves are not yet unfolded.

The burst of floral activity which started in April reaches a crescendo in late May when 171 species were observed to bloom for the first time during this month. From February to June, blooming is geared by abundant moisture and favorable insolation to a geometric progression. From June until November, the procession of bloom slips into an arithmetic decline (Fig. 2).

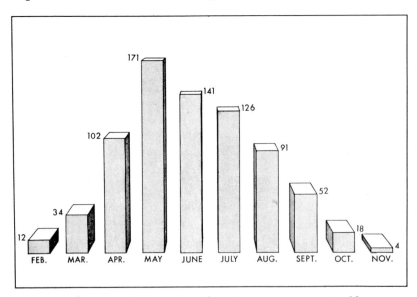

2. A nearly geometric progression of species coming into spring bloom at Giant City State Park and a slow arithmetical monthly decline through summer and autumn.

Shortly before and following the summer solstice, the blooming periods become shortened, but their variability from year to year, by comparison with their average dates, shows a higher degree of reliability for flowering dates than those of spring (Leopold and Jones, 1945).

Moisture, abundant in earlier months, is now removed by dryness of July weather. Days, like candles, are burned shorter. These conditions induce floral and growth changes. Only such hardy and deep-rooted plants as sunflowers, coreopsis, wild bergamot, and rosinweed are seen with regularity. Many plants now commonly seen show prairie affinity. Dry weather continues into the fall aspect, the appearance of which is forecast by the arching flower clusters of early blooming goldenrods.

The brightly colored cover of spring and summer now is exchanged for August's yellows, copper, and dull green. The long green season of food-making labor and storage gives way to gorgeous color and fruition. Among the herbs, a last burst of color is made by fruiting and flowering heads of lespedeza, whites of spurges, yellows of evening primroses, and purples, blues, and whites of asters growing thickly on low ground and open roadside areas.

The drought conditions of September, greater night and day temperature fluctuations, clear bright days, and sugars trapped in cells suddenly grown old, all combine to give rich autumnal coloration. First to turn are sassafras and sumac, followed by dogwood and sour gum, with colors of orange-brown and flaming red. Now maples, ashes, and hickories burst forth with yellows while the oaks are only sere save the white oak which turns a dark red. Sweet gum runs a gamut of color—from yellow through red and even almost purple. All colors are intense, but soon fade to leave behind ripening fruits on bare branches.

Rattling cornstalks pursued by husky winds flee the desolation of fields now stripped of their golden saving of sunlight. An endless vault of blue sky and land spilled over and spattered with yellow is pushed and bullied from the scene. Fingers of frost clutch and surround turgid persimmons and gently release them from their branches; they fall to a cushion of bluegrass and flatten with the impact, or softer ones are speared through by stiff saber-like blades. We may be sure there is not a trace of pucker in these. Scats on a sunny ledge tell us foxes also have found the persimmons ripe.

Beech trees reach with light gray filigreed branches into the darker fish-scale and mother-of-pearl sky. Intermittent slants of sunlight streak through between cloud layers and in the brief sunlit interval nature comes to life again. The sight of wild bees winging to a hole in a beech tree is our reward for patience and observation. Patience could be taken for granted in a windless valley embraced with occasional warm hugs; we are not in a hurry. Who knows how great the store of liquid sunshine lying in the hollow of that great gray bole? Who knows it is there aside the bee and me? Some animal knows, we are sure, for he writes this knowledge for us to read in the path he has worn around the base of the tree in his effort to get inside.

A history of northern ancestry is a chapter in this almanac of the seasons. As we cross a spring-branch we note the inseparable association of cold water and alder growth. Should we miss this hint of its northern ancestry, this species tells us again in its early blooming date. Catkins are conspicuously displayed as early as March when it is still too rigorous a climate for all but the most hardy. The great white bear sedge is another northerner; witness its wide green leaves of the winter season. Other hardy plants noted for their green leaves are easily spotted among the brown leaves of the forest floor. These hardy plants include the waterleaf, avens, partridge-berry, christmas fern, heal-all, bedstraw, wild onion, violet, wind flower, putty-root orchid, wild leek, and others. Most other plant life after vigorous growth and effort towards fruition now retreats underground or into seclusion of buds on barren branches.

These buds, standing in modest array, command hardly a second look. They have come to be known as winter buds, as if they made their sudden, almost spontaneous, appearance then. In reality, as last year's leaves were unfolding in April and May, these buds, now visible during winter, were being formed and filled with the verdure of a promised but yet unborn summer.

Land Antiquity

Some more important geological events having bearing upon plant distribution are previous land submergences, uplifts, deformations, and glacial advances and retreats. In view of the magnitude and importance of such geological events upon development of vegetation, it seems advisable to examine the geological history of our area. (Stewart A. Weller's paper, "The Making of Southern Illinois," has been especially helpful in the preparation of this section.)

Time stands still in a wooded valley bestrode by massive bluffs of limestone or sandstone. The oldest part of our landscape are these rocks. They came up out of the sea whence they were laid down. These sedimentary rocks could not have been formed except for erosion of higher land areas. The erosional cycle of youthful topography, maturity, and old age to base level and uplift again has been enacted countless times, and the geological erosion of the present is still referring sediments to the sea for deposition. Many present marine forms now lying dead upon the ocean floor are being covered slowly by fine blankets of limy or sandy deposition. Subsequent replacement of softer body parts by minerals such as silica or lime produces fossils. Thus the newest layers are now being laid down, and the history of present day marine organisms is being recorded in these strata as they were in those of the past.

Over a period of time from storing newspapers or magazines, the oldest ones are to be found at the bottom of the pile and more recent ones near the top. Oldest rings in a tree trunk are similarly near the center with newer ones to the outside. Oldest leaves from litter of the forest floor are likewise to be found deepest with successively newer ones on top. This same type of chronology may be found in the earth's rock layers.

Geological time is reckoned in millions and hundreds of millions of years. One of the most useful methods of dating rocks from different layers has been the rate of decay of certain radioactive minerals found in them. The earth itself is, by this method, estimated to be at least three-and-one-half billion years old and it is suspected of being even older. It is thus understandable that in the study of earth history a million years is as but a day or so. Forces of nature observed to be operating in the present may seem to be of small importance, but over the span of geological time, the effects are profound.

The fossil record bound in the rock layers of the earth's crust has been used by geologists as a basis for dividing the whole of geological time into eras. These divisions are designated as the *Proterozoic* during which primitive invertebrate forms predominated; the *Paleozoic*, spanning the interval between Proterozoic and a "middle period" known as the *Mesozoic;* and the *Cenozoic* in which the forms of life were more like those of the present.

It is only for convenience of study and ease of reference that a continuous thing like time is divided into eras. For greater detail, each era has been sub-divided into lesser divisions, called periods (Table 3). Often a period name may be for a geographic area where rocks representing the period are common and easily observed. Rocks bearing fossils of organisms existing around three hundred million years ago were first discovered in Devon, England; hence, these rocks are known as *Devonian*. Later were the layers from the Mississippi valley; hence, these are from the *Mississippian* period. Still later, at about two hundred and fifty million years, is the *Pennsylvanian* period, named for the locality of its discovery. These three periods, the Devonian, Mississippian, and Pennsylvanian are divisions of the late Paleozoic era.

Rocks of the late Paleozoic era are found easily in southern Illinois today. Devonian rocks are found exposed in southwestern Illinois (Jackson, Union, and Alexander Counties) and in southeastern Illinois (Hardin County). In Hardin County the Devonian strata are exposed in a structure known as Hicks Dome. The dome is assumed to have been a giant bubble in the earth's crust, formed by gases and igneous rock, which subsequently cooled and solidified. It is supposed

3. A geological chronology of earth's history

ERA	PERIOD	EPOCH	Character of Biota
Cenozoic (75)	Quaternary (1)	Recent	Rise of civilization
		Pleistocene or Ice Age	Extinctions and migrations
	Tertiary (74)	Pliocene	Numerous herbaceous types and grazing animals
		Miocene	Peak carnivore populations
		Oligocene	Increased mammalian types, warmth-loving vegetation
		Eocene	Rich angiosperm forests, browsing herbivores
Mesozoic (130)	Cretaceous (60)		Rise of Angiosperms
	Jurassic (30)		Widespread conifers and cycads, first birds, modern fishes
	Triassic (40)		Earliest mammals, plant life mostly cycadophytes
Paleozoic (275)	Permian (25)		Few plant types, many reptiles
	Pennsylvanian (25)		Coal forests and swamps, great rise of reptiles, early insects
	Mississippian (25)		Lycopods, horsetails, seed ferns
	Devonian (45)		First extensive land flora of vascular plants
	Silurian (35)		First fossils of land plants
	Ordovician (65)		Aquatic life
	Cambrian (80)		Trilobites, other marine invertebrates
Proterozoic (900)	Pre-Cambrian		?

Figures are estimated durations in millions of years; adapted from Sinnott and Wilson, 1955, p. 311.

that the Hicks region may have, at an early time, been much higher than it is today. It has been eroded to an extent that an old neck plug of the volcano may be located (Bonnell, 1946). Devonian rocks completely encircle the Hicks Dome outcrop. They are continuous, but underlie other strata across the southern tip (Fig. 3). They outcrop again in Union County and southern Jackson County at the Devil's Bake Oven where these strata are tilted at an angle of about twenty-six degrees. Devonian rocks are limestones containing shells of marine animals. By fossil and rock correlations the Devonian sea is reckoned

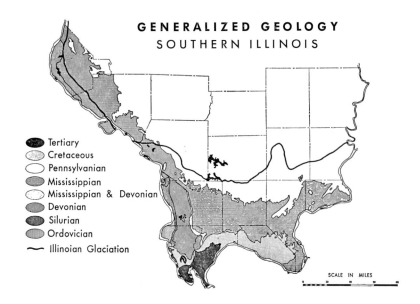

3. Geologic map of the Southern Illinois area. (Adapted from J. Marvin Weller, *et al.*, Division of State Geological Survey, State of Illinois, 1953, second printing.)

to have extended as far north as Canada and New York State and the Allegheny Mountains. It extended as far west as the Mississippi River of the present except in southern Missouri where it extended farther west. A broad connection with the ocean doubtless occurred in the south. At least two islands of considerable size existed in this sea. One extended northward from the present location of Cincinnati, the other in central Tennessee near the present location of Nashville.

Mississippi limestones, because they are younger than Devonian strata, are found as an outer encircling formation of Hicks Dome in Hardin County. Still younger rocks encircle the Mississippian strata, etc. Limestone bluffs occur from Rosiclare to Cave-in-Rock. Abundant limestone in the area, though covered, is indicated by the presence of numerous sink-hole ponds and lakes.

Vegetation made its appearance in the watery realm of the Ordovician and Silurian seas. Soon an uplift of land left muddy shores and shallows. Nature's cautious experiments at life on land proved successful and the new environment was rapidly exploited. The gains made by this vanguard of vegetation, made up of primitive ferns and their allies, were further consolidated in the Devonian period and a few "seed ferns" are known to have existed toward the end of this period.

Toward the close of the middle Devonian, great crustal dislocations were thought to have occurred in this area causing southeastern Missouri and possibly part of southern Illinois to emerge from the sea. Toward late Devonian, southern Illinois became submerged again as a part of an inland sea whose extent was from the Appalachians north to Lake Erie, west to Oklahoma, and south to southern Tennessee. Deposition of about four hundred feet of a black shale indicates a long tenure for this encroachment by the sea. Southern Illinois was to be uplifted, eroded, and submerged again in a sea of the Mississippian period.

More extensive were these Mississippian seas; they reached westward to Montana and southward to New Mexico. Eastward and southward their waters lapped on Alabama and connected with the ocean further south. Their touch northward was felt by southern Wisconsin. Still eastward they were contained by the large land area known as Appalachia. Toward the close of Mississippian time southern Illinois was raised once more and remained out of the water for a lengthy period, but was engulfed yet another time. The rather pure limestones of the earlier submergences gave way to layers mixed with sand and fine mud, identified as the Chester series. During the Chester epoch, southern Illinois was raised and lowered possibly several times. Because of the mixing of sands with lime-mud it is thought that the existing land area was actively eroded and the shoreline was shifted possibly several times. The whole southern portion of North America became a land area with the subsidence of the Mississippian sea.

The Pennsylvanian period which followed initiated a long period of erosion. Sands and silts flowed into depressions to fill them or make them more shallow. These beneficient contours and a particular degree of climatic favor produced a wealth of plant growth. In many local areas were *Calamites, Sigillaria,* and *Lepidodendron.* These were giant horsetails and clubmosses. Through their demise, fall, and fill of darkened swamp waters, and their coverage by the silts of time, this vegetable material became compressed into rich beds of coal. Indeed, because of the great accumulations of these coal measures, the two periods of Mississippian and Pennsylvanian have been combined into one which has been called the Carboniferous. Materials of the sandstone presently forming the crest of the "Ozark" or Shawneetown Ridge in southern Illinois were washed into the sea during the Pennsylvanian period.

The present contrast in relief between the Illinois Ozarks and surrounding areas in the state is due to a difference in rock hardness. Those to the north were not so hard. Our area may possibly have been uplifted to some extent with the adjacent hilly land of the Missouri Ozarks. Greater relief of our area to proximal areas of Illinois also has been linked with deep-seated volcanic activity. Igneous dikes have been found in coal seams at Harrisburg and also in the Kentucky coal field area. These igneous intrusions have been associated with Appalachian Mountain disturbance.

After Pennsylvanian strata were formed, there were re-adjustments in the earth's crust in this region. Bowing, breaking, and overlapping of these layered rocks produced extensive fault systems in southern Illinois and western Kentucky. Slight deformations still are occurring farther south along some of these old fault lines. Area residents will recall slight tremors during the late nineteen fifties. Nearly one hundred and fifty years ago a severe earthquake was felt throughout the area. Centered around New Madrid, Missouri, this disturbance was from re-adjustments of greater than usual magnitude in this area.

Strata of late *Mesozoic* Age (Cretaceous period) in the extreme southern portion of Illinois have had no visible deformations (Fig. 3). It was during the *Cretaceous* period that the modern Angiosperm flora had its beginning. Woody flowering plants appeared gradually and began their deployment over the land areas to become the dominant constituents of the vegetation.

The climate of the early Tertiary was warm and in the Mississippi embayment shallow swamps were abundant. The embayment, being continuous with the coastal plain, harbored many psammophilous plants and aquatics. Some plants whose ancestors inhabited these low swampy forests are bald cypress (*Taxodium distichum*), swamp cottonwood (*Populus heterophylla*), water hickory (*Carya aquatica*), swamp red maple (*Acer drummondii*), tupelo (*Nyssa aquatica*), pumpkin ash (*Fraxinus profunda*), water elm (*Planera aquatica*), overcup oak (*Quercus lyrata*), and water locust (*Gleditsia aquatica*). Other woody plants with strong southern affinity include pawpaw (*Asimina triloba*), redbud (*Cercis canadensis*), cane (*Arundinaria gigantea*), spanish oak (*Quercus falcata*), deciduous holly (*Ilex decidua*), and carolina buckthorn (*Rhamnus caroliniana*).

Other modern genera such as magnolia, persimmon (*Diospyros*), sassafras (*Sassafras albidum*), sycamore (*Platanus*), hackberry (*Cel-*

4. A few plants related generically to those existing when dinosaurs roamed in the late Mesozoic (Cretaceous). Upper left—magnolia; upper right—persimmon (*Diospyros*); left center—sassafras; right center—sycamore (*Platanus*); lower left—beech (*Fagus*); lower right—hackberry (*Celtis*).

tis) (Fig. 4), as well as dogwood (*Cornus*), walnut (*Juglans*) and yellowwood (*Cladrastis*) claim generic relations in the late Mesozoic (Cretaceous).

"While there may have been differentiation of floristic types in the Cretaceous, correlated with climatic conditions, the first of the great vegetation segregations which is still of importance in our region began at the close of the Cretaceous with the uplift of the Cordilleran complex of mountains. Intercepting the moisture-laden winds from the Pacific and restricting the rainfall of the lands immediately east of them to moisture derived from the Gulf of Mexico, the elevation of these mountains led to the development of semi-arid conditions over the Great Plains, which soon had an effect on the character of the vegetation. The result was the grassland type which still prevails in that region" (Gleason, 1923). Prairie vegetation existing south of the limits of glaciation is a lineal descendent of the climatic prairies of the Tertiary period. Two kinds of prairies exist in southern Illinois at present—those of unglaciated areas which are oldest and the more recent prairies of the glaciated areas.

In the Cenozoic era the outstanding geological epoch was the *Pleistocene* or "Ice Age." With the passing of many millions of years, ice and snow accumulated in the north. Sometime within the past million years, the cold colossus of continental glaciation began its slow, but persistent movement southward. Bluish-white, crystalline, cutting in coldness, and awesome in its thickness which was often sometimes as great as a mile, it relentlessly descended from Keewatin and Labrador.

Climatic conditions prevailing during the reign of the ice sheets were no doubt similar to that which sustains active glaciers today. (This section on glacial conditions has been paraphrased and abridged from the paper, "Scenes in Ohio During the Last Ice Age" by Richard P. Goldthwait, *The Ohio Journal of Science* 59[4]:193. July, 1959.) Discovery of spruce logs in Ohio, covered during the last (Wisconsin) glaciation suggests the mean annual temperature then was 23° F. or some 30° colder than today in this area, and that for central Ohio the July temperature was about 20° colder than today. Heavier precipitation with much snow was a necessary condition to nourish the glacier and foster its southward movement. Active large glaciers today receive 60 to 110 inches of snowfall on their surface annually. Northern areas today which support spruce trees with trunk sizes correlating with relict bogs in Ohio exist as far north as 54° N. latitude where mean temperature is 23° F., the mean daily temperature in January is 11° F., and mean daily July temperature is 55° F.

Ice movement is due to slipping at the base or along the sole of the glacier. Water in this zone of contact between earth and glacier lubricates the base for slippage because of water being of greater density than ice. Thus where water issued from or existed in crevasses it exerted an upward or hydrostatic force. This melting at the base is assumed to be continuous, even through the winter, for it is known that rivers flow in this season from under Greenland ice caps today.

Down the Lake Michigan basin and over the broad lowlands of Illinois, the glacier rode, gouging, scouring, grinding and chiseling — it planed and smoothed the hills and filled the valleys and depressions with its load. Water from the melting ice was milky in color and copious in its escape from the edges of the ice. Melting occurred in summer periods while a heavy nourishment of snow accumulation occurred in winter to drive the glacier forward. Similar to the manner of a person's advancement up a sand hill was the progress of the glacier. Forward progress was "two steps" in winter while "sliding back" one in summer. Finally the ice began a decay at the southernmost reaches of the glacier. Melting water poured into holes several miles back from its border to create natural tunnels in the ice. As water from these tunnels headed down valley lowlands, glacial till, sand, and gravel were spread over the land as velocity was checked. The direction of river flow was often confused by deposition of heavier materials and subsequent fill and overflow of channels. Braided stream patterns were created as many millions of gallons of melt water were produced on the warmest days.

Modern glaciers are caused to retreat in Iceland by mean annual temperature of 39° F. and summer mean temperature of 51° F. A summer average of 56° F. induces a retreat of glaciers at Juneau, Alaska. Thus Goldthwait (1959) postulates that when their earlier continental glaciations were beset with summer mean temperature of about 56° F., they must have released their cold grip of the land.

Migrating plants making a successful forced march in the threatening advance and ominous shadow of the nearby ice sheet have lingered in southern areas where physical diversity has provided some compensation of factor or factor complexes. The migration was achieved through seeds and fruits being scattered by winds, water, and other agents. Even the ice itself deposited seeds and fruits at southern stations as the ice advance was halted by a warming period. Several species today are clustered near the former glacial border and show a considerable disjunction between these locations and present-day northern stations. Plants with less mobile seeds and fruits were ground to oblivion by the monstrous thickness and weight of ice.

The earliest of four glacial advances, the Nebraskan, did not touch southern Illinois. As the Nebraskan ice receded northward, there was a warm inter-glacial period followed by another glaciation known as the Kansan. Ice from this glacier came south to cover about one-half of Randolph county (Fig. 5). The third ice sheet, the Illinoian, covered our state except for a small driftless area in northwestern Illinois and the southern tip. This period of glaciation represented the southermost penetration of any of the four glacial advances. It reached as far south in Illinois as the Shawneetown Ridge before it overreached the climatic coldness to sustain its march. The ridge or escarpment was a barrier holding back the dying glacier (Fig. 5). The fourth ice sheet, the Wisconsin glaciation, was most recent. It is thought to have begun its recession only eleven thousand years ago. It covered only about the northeast one-third of the state. The colder elements prevailing during these glacial times initiated migrations of plants and animals southward. The drainage changes, the grinding and pulverizing of rock materials, the drift left by melting waters, and the deposition of loess all must have had profound influences upon our flora.

GLACIATION IN ILLINOIS

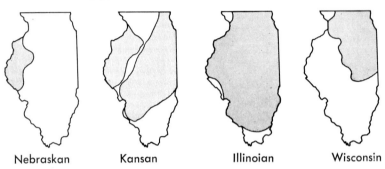

Nebraskan   Kansan   Illinoian   Wisconsin

5. Maps showing advance of the four glaciers into Illinois. (Adapted from Ill. Academy of Science Field Trip Guide Leaflet No. 61, by Ill. State Geological Survey, Urbana, 1961.)

Vegetational History and Floristics

A blue haze hangs softly over undulations of worn sandstone hills in the area known as the Illinois Ozarks. The blue haze and the general accordance of summit levels remind one of the greater mountains eastward over our continent (Fig. 6). The Shawnee Hills are but miniatures of the Appalachians in both height and area, and are not nearly as old.

Deciduous forest vegetation today is descended from the Arcto-tertiary forest which, according to the fossil record, was made up of both gymnosperms and angiosperms. Moreover, this forest of the Tertiary was composed of a mixture of temperate genera and other genera which today are regarded as tropical and subtropical. It is also indicated there was little latitudinal differentiation as far north as seventy degrees (Gleason, 1923).

Forest vegetation during the Pleistocene or ice age was forced southward and prevailing forms contracted into a refuge area centered in the southern Appalachians of eastern Kentucky, Tennessee, and western North Carolina. It was from this strong-hold that the vast deciduous forest spread and multiplied into countless communities of trees (Braun, 1950).

With climatic shifts, regional vegetation and flora are affected. Over a long period, as the climatic pendulum swings back again, some small stands or single individuals remain behind in compensating areas. These leftover individuals or stands of vegetation are relicts. They stand alone in an area where they were formerly abundant. An adjoining or sometimes a more remote place now may have a greater

6. The Shawnee Hills, heavily forested and with blue haze hovering over them, suggest a miniature of the Appalachians. Horseshoe Bluff in the Shawnee National Forest overlooks lowland in Jackson County.

number of species or development of a particular type of vegetation. Relicts are frequent along the glacial border. Some factors important in the persistence of northern plants near the glacial border are springs where cold water seepages compensate for the bog habitat. Water willow (*Decodon verticillatus*) has found its northern requirements met in this situation. Shade of north-facing slopes affords a lower temperature and a higher humidity in summer when such factors are critical in the existence of plants of more northern character. Finding these haunts favorable are partridge-berry (*Mitchella repens*), bog moss (*Sphagnum* spp.), and ground pine (*Lycopodium complanatum* var. *flabelliforme*), while the wild leek (*Allium tricoccum*) is favored by the drainage of cold air into moist valley bottoms.

Plant geographers use, among other criteria, the greatest number of genera and species, the place of their greatest frequency of occurrence, and the place of occurrence of individuals of greatest size to determine the center of area (Cain, 1944). Over 130 native tree species are to be found in the Great Smoky Mountain National Park area. Many of these species have high frequency of occurrence and reach extreme size for their respective kinds. Thus it may be that this forest of the southern Appalachian region is the progenitor of other deciduous forest types including those of southern Illinois.

The over-all relief of the Illinois Ozark area is only about eight hundred feet and the average annual rainfall is less than half that of the Smokies. In spite of our lesser rainfall and junior stature of our mountains, we make no concessions to the Appalachian area in floristic richness. Indeed we may, in a way, boast superiority; for on slightly over two square miles at Giant City State Park in Jackson County have been collected over 820 species of vascular plants (ferns, their allies, and seed plants).

We cannot claim several biotic zones with the upper altitudinal one being climatically equivalent to far away northern regions. We have no spruce or fir trees, no rose bay, no catawba rhododendron, no trailing arbutus nor other entities of exquisite reknown. We may, however, proudly point to many plants in our area which have floristic affinity with the Appalachian area. We have, for example, chestnut oak (*Quercus prinus*), pinxter flower (*Rhododendron nudiflorum*), filmy fern (*Trichomanes boschianum*), Bradley's spleenwort (*Asplenium bradleyi*), cordate black cohosh (*Cimicifuga cordifolia*), alum-

7. [left] Alumroot (*Heuchera parviflora*) in a moist rock-wall crevice. [right] Stonecrop (*Sedum telephioides*) on a north-facing rock ledge at Belle Smith Springs in Pope County.

root (*Heuchera parviflora*) (Figs. 7 and 8), stonecrop (*Sedum telephioides*) (Figs. 7 and 8), hedge nettle (*Stachys clingmanii*), water leaf (*Hydrophyllum macrophyllum*), tulip tree (*Liriodendron tulipi-*

9. [left] Harvey's buttercup (*Ranunculus harveyi*) grows in profusion on the gentle sloping upland at Piney Creek in Randolph County. [right] Flower-of-an-hour (*Talinum calycinum*) in thin mineral soil of a sandstone ledge at Rock Castle Creek in Randolph County.

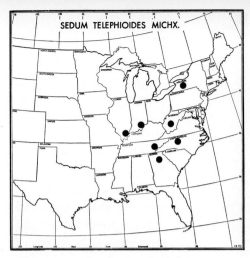

8. [left] Alumroot shows Appalachian affinity in its distribution. [right] Stonecrop shows Appalachian affinity in its distribution. (These and other maps showing species distribution are reproduced with permission of the George F. Cram Co., Indianapolis, Indiana.)

fera), cucumber magnolia (*Magnolia acuminata*), and silver bell (*Halesia carolina*).

Another interesting feature of the southern Illinois flora is the presence of Missouri and Arkansas Ozark species in the southwestern counties. These species have managed to cross the Mississippi River and have made themselves rare members of the Illinois flora. Numbered among these are Butler's quillwort (*Isoetes butleri*), flower-of-an-hour (*Talinum calycinum*) (Figs. 9 and 10), shortleaf pine (*Pinus echinata*), Harvey's buttercup (*Ranunculus harveyi*) (Figs. 9 and 10), Missouri primrose (*Oenothera missouriensis*), bedstraw

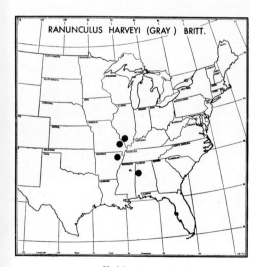
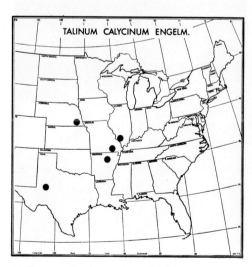

10. [left] Harvey's buttercup is essentially an Ozark species. [right] Flower-of-an-hour shows, in its distribution, an Ozark or western affinity.

(*Galium virgatum*), ozark beards-tongue (*Penstemon arkansanus*), slender heliotrope (*Heliotropium tenellum*), ozark goldenrod (*Solidago strigosa*), ozark coneflower (*Rudbeckia missouriense*), and numerous sedges (*Carex austrina, C. emmonsii,* and *C. physorhyncha*).

The south-central United States (Ozarks) have been presented by Palmer and Steyermark (1935) as a possible center of origin for many species. They point out that many species have persisted there since pre-glacial times and that the area has been continuously available for occupancy by plants since late Paleozoic. This proposition is supported with the idea that extensive migrations into the Ozarks from the east would have been unlikely after the development of the wide floodplains of the Mississippi River. The river is viewed as having been too formidable a barrier to migration by many upland species.

Being geographically situated between the Appalachians and the Ozarks, our area owns a natural relationship to both and thus becomes a giant link in nature's floristic gradient. A general reduction in size of individual plants and their numbers occurs in the spread of species to the West. A climatic gradient condition from moist to dry is the

11. [left] Bishop's cap (*Mitella diphylla*) on a shaded, moist, north-facing sandstone rock at Giant City State Park, Jackson County. [right] Ground pine (*Lycopodium complanatum* var. *flabelliforme*) trails over a north-facing rock shelf at Indian Kitchen on Lusk Creek in Pope County.

underlying cause. There lie to the West, to the North, or to the South, floral assemblies of different origin and nature. The vegetational character of our area owes much to its geographic location and geological history. Under compulsion of continental glaciation, northern plants were laid at our doorstep. Among these may be mentioned the harebell (*Campanula intercedens*), nannyberry (*Viburnum lentago*), bishop's cap (*Mitella diphylla*) (Figs. 11 and 12), sedges (*Carex substricta, C. media, C. tetanica,* and *C. projecta*), ground pine (*Lycopodium complanatum* var. *flabelliforme*) (Figs. 11 and 12), staghorn sumac (*Rhus typhina*), and verticillate holly (*Ilex verticillata*).

Beyond the waves of the inland sea were several ancient Atlantic Coast species. Among these are Virginia willow (*Itea virginica*), southern buckthorn (*Bumelia lycioides*), storax (*Styrax americana*), American featherfoil (*Hottonia inflata*), sedges (*Carex louisianica, C. oxylepis, C. flaccosperma,* and *C. decomposita*), tupelo (*Nyssa aquatica*), bald cypress (*Taxodium distichum*), sponge plant (*Limnobium spongia*), swamp iris (*Iris fulva*) (Figs. 13 and 14), spider lily (*Hymenocallis occidentalis*) (Figs. 13 and 14), *Triadenum tubulosum, Hydrolea affinis,* and *Glyceria arkansana*.

Mountain building in the West (rise of the Rockies) and subsequent climatic change interrupted the far flung deciduous forest and led to the formation of our present mid-continental grasslands or prairies. Extensive migrations of these prairies have occurred in the

 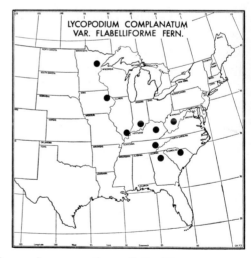

12. [left] The distribution of bishop's cap shows a northern affinity. The southern stations are at higher elevations, thus the temperature of more northern latitudes is compensated. [right] Ground pine is a plant with northern affinity. It ranges beyond 60 degrees north latitude.

13. [left] Swamp iris (*Iris fulva*), whose bloom, dark brick-red in color, may be seen in the LaRue Swamp in Union County. [right] Spider lily (*Hymenocallis occidentalis*) is known from a few swampy woods in Southern Illinois.

past under the compulsion of climatic change. Scattered grassland communities may now exist within the forest areas. These relicts are known locally as hill prairies. Only miniatures in size, they are good replicas of the larger prairies westward, and reveal their lineage by the presence of western insects or other forms which also made the migration.

The hill prairies present an array of herbs which in their seasonal aspects resemble myriads of jewels in the grass. A scarcity of taller grasses is observed at once and conversely there is a greater amount of little bluestem (*Andropogon scoparius*) and side-oats grama (*Bouteloua curtipendula*). Most outstanding of herbs are puccoon (*Lithospermum canescens*) of the spring aspect, the coneflower (*Echinacea pallida*), and the purple and white prairie clovers (*Petalostemum purpureum* and *P. candidum*) of summer aspect.

Prairie vegetation generally is not present in southern Illinois on glaciated plains adjacent to the border of Illinoian glaciation. It is present in extensive areas of Wisconsin glaciation where it developed rapidly upon the recession of the last glacial period. These are portions of True Prairie and dominated by tall grasses such as big bluestem (*Andropogon gerardii*), Indian grass (*Sorghastrum nutans*), and switchgrass (*Panicum virgatum*).

The southern Illinois area, by its antiquity and location with respect to other floristic centers (Fig. 15), and by the vicissitudes of crustal upheavals, rise and recession of seas, glaciation and climatic change,

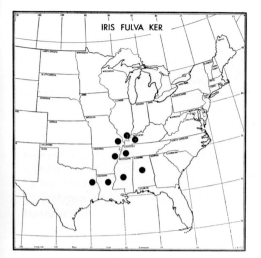 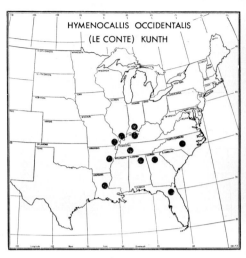

14. [left] Swamp iris has a southern or Mississippi-embayment distribution pattern. [right] Spider lily distribution suggests a Coastal-Plain affinity.

has become a "floristic melting pot." The monstrous forces of the past may be credited with the major role in our floristic potpourri, but a force of considerable import which continues into the present is the rivers. Our area is situated between the lower Wabash and Ohio on the east and the Mississippi on the west. Species which have used the valleys and continue to use the water highway for transport are legion.

Rensing (1957), in determining the floristic affinities of 1,501 vascular plants in the Illinois Ozark area, checked the geographic ranges of all plants from several existing manuals or floras. After establishing the distribution of species, each one's occurrence then was related to whether it was an intraneous or extraneous species (Table 4). This follows the classification of Cain (1930) which was earlier proposed by Cowles (1929).

Intraneous species are those indigenous to southern Illinois and not near the limits of their distribution. The intraneous category may further be subdivided into: (1) plants of Eastern United States, whose members range generally throughout the entire eastern half of North America with the major center of distribution being east of southern Illinois; (2) plants of Central United States with southern Illinois being near their center of distribution; (3) plants of Southeastern United States, some of which have advanced northward and often westward beyond the geographic limits of southern Illinois.

Extraneous plants are those which are at or near the limits of their

VEGETATION REGIONS

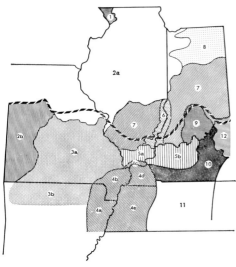

1. DRIFTLESS AREAS
2. PRAIRIE
   a. GLACIATED
   b. UNGLACIATED
3. OZARKS
   a. MISSOURI
   b. ARKANSAS
4. MISSISSIPPI EMBAYMENT
   a. EAST ARKANSAS
   b. SOUTHEAST MISSOURI LOWLANDS
   c. TERTIARY DIVISION, ILLINOIS
   d. WESTERN KENTUCKY
   e. WESTERN TENNESSEE
5. SHAWNEE SECTION
   a. ILLINOIS OZARKS
   b. WESTERN COAL FIELDS
6. WABASH LOWLAND
7. TILL PLAIN
8. LAKE AREA
9. INDIANA OAK-CHESTNUT UPLAND
10. MISSISSIPPI PLATEAU
11. MIDDLE TENNESSEE
12. APPALACHIAN
    ▰▰ GLACIAL BORDER

15. Vegetation regions of the Central-States area. Regions shown are based on physiographic divisions. (Map adapted from several sources—Palmer and Steyermark, 1935; Jones, 1950; Deam, 1940; and Braun, 1943.)

distribution or range in southern Illinois. This has been arbitrarily set at one hundred miles of the border of the southernmost twelve-county area. Sub-groups are based upon the extent of range to the southeastern, southern, southwestern (Ozark), central United States, or northern North America. Thus the first three floristic groups (Table 4; Groups 4, 5, and 6) are northern. Groups 4 and 5 include those species mainly distributed in northeastern and northwestern United States, respectively. Group 6 contains species growing in Canada and extending across the continent. Groups 7, 8, and 9 are southern, with major distribution being to the southeast, south, and southwest, respectively. Introduced species are native to another continent or another part of North America, other than the continental United States. Some cultivated plants have escaped and subsequently have become naturalized.

Conspicuous facts revealed in this survey are the low percentage of species showing affinity to the north (less than 10 per cent) and the relatively high percentage of species showing affinity to the east and southeast. If one were to include the category of plants that range throughout the eastern United States and to assume these species to be geographically centered east of southern Illinois, the total having an affinity to the east or south would be seventy-one per cent of the flora (Rensing, 1957). This would support the observation that the vegetation of the Hill Section (including the Illinois Ozarks) is distinguished by its strong Appalachian element (Braun, 1950). This is rather to be expected for the physical diversity of southern Illinois affords many favorable habitats in the numerous limestone sinks, hollows, fresh and salt water springs, bluff overhangs, natural bridges, clear running rock-bottomed streams, and rocky uplands. A directly proportional number of biological entities to habitat variation is claimed. The vegetation of the Appalachians is rich in number of species and its development is luxurious. Because the Appalachians and adjacent dissected uplands present a continuity of habitat and provide an outlet to the north and west, the eastern element in the Illinois Ozarks is pronounced.

Considering the extraneous plants as a complete segment of the flora, it is shown that better than one in every five species (327 in all) reaches the edge of its range here. These extraneous members extend in all directions: north, east, south, and west (Table 4).

4. *Geographical affinities of southern Illinois plants*

GEOGRAPHIC AREAS	No. of Species	Per Cent of Total
INTRANEOUS	964	63.5
Eastern United States	710	46.5
Central United States	141	9.5
Southeastern United States	113	7.5
EXTRANEOUS	327	22.5
Northeastern United States	91	6.0
Northwestern United States	40	2.5
Canadian Transcontinental	13	1.0
Southeastern United States	100	7.0
Southern United States	56	4.0
Southwestern United States	27	2.0
INTRODUCED	210	14.0
Total	1501	100.0

After Rensing, 1957.

Extraneous Species

CANADIAN TRANSCONTINENTAL

Alopecurus aequalis Sobol.
Arctium minus (Hill) Bernh.
Bidens cernua L.
Carex bebbii Olney
Eleocharis palustris (L.) R. & S.

Equisetum arvense L.
Muhlenbergia mexicana (L.) Trin.
Najas flexilis (Willd.) R. & S.
Scirpus fluviatilis (Torr.) Gray
Spartina pectinata Link

NORTHWESTERN UNITED STATES

Allium stellatum Ker
Anemone canadensis L.
Apocynum sibiricum Jacq.
Aristida longespica Poir.
Artemisia biennis Willd.

Artemisia gnaphalodes Nutt.
Asarum canadense L. var. reflexum (Bickn.) Robins.
Carex gravida Bailey
Carex muskingumensis Schwein.
Cheilanthes feei Moore

Coreopsis palmata Nutt.
Cornus stolonifera Michx.
Cyperus schweinitzii Torr.
Diarrhena americana Beauv.
Echinochloa occidentalis Wieg.

Gerardia aspera Dougl.
Gerardia grandiflora Benth. var. pulchra (Pennell) Fern.
Heuchera richardsonii R. Br. var. affinis Rosend., Butt. & Lak.
Juncus interior Wieg.
Koeleria cristata (L.) Pers.

Oenothera biennis L. var. canescens T. & G.

Oenothera pilosella Raf.
Ophioglossum vulgatum L.
Panicum leibergii (Vasey) Scribn.
Pilea opaca (Lunell) Greene

Potentilla paradoxa Nutt.
Pyrus coronaria L.
Pyrus ioensis (Wood) Bailey
Rhus aromatica Ait. var. illinoensis (Greene) Fern.
Robinia hispida L.

Rumex mexicanus Meisn.
Salix amygdaloides Anders.
Salix cordata Muhl.
Salix interior Rowlee
Smilax ecirrhata (Engelm.) S. Wats.

Solidago graminifolia (L.) Salisb. var. media (Greene) Harris
Sparganium eurycarpum Engelm.
Tephrosia virginiana (L.) Pers. var. holosericea (Nutt.) T. & G.
Thalictrum dasycarpum Fisch. & Lall.
Vernonia fasciculata Michx.

Acnida subnuda (S. Wat.) Standl.
Agropyron repens (L.) Beauv.
Allium tricoccum Ait.
Antennaria parlinii Fern.
Asclepias exaltata L.

Asclepias verticillata L.
Aster simplex Willd.
Aster tradescanti L.
Athyrium filix-femina (L.) Roth var. rubellum Gilb.
Bidens coronata Britt.

Bidens polylepis Blake var. retrorsa Sherff
Boltonia asteroides (L.) L'Her.
Botrychium dissectum Spreng.
Bulbostylis capillaris (L.) C. B. Clarke
Bromus ciliatus L.

Carex annectens Bickn.
Carex brachyglossa Mack.
Carex careyana Torr.
Carex cephaloidea Dewey
Carex cristatella Britt.

Carex emmonsii Dewey
Carex grayii Carey
Carex hirtifolia Mack.
Carex lacustris Willd.
Carex platyphylla Carey
Carex styloflexa Buckl.
Carex umbellata Schk.
Carex woodii Dewey
Cassia hebecarpa Fern.
Chelone glabra L. var. linifolia Coleman
Comandra umbellata (L.) Nutt.

Cornus racemosa Lam.
Cyperus engelmannii Steud.

Danthonia spicata (L.) Beauv.
Decodon verticillatus (L.) Ell.
Eleocharis intermedia (Muhl.) Schultes

Eleocharis smallii Britt.
Erechtites hieracifolia (L.) Raf.
Fagus grandifolia Ehrh.
Gerardia paupercula (Gray) Britt.
Gerardia tenuifolia Vahl var. parviflora Nutt.

Geum virginianum L.
Helenium autumnale L.
Heliopsis helianthoides (L.) Sweet
Hieracium scabrum Michx.
Holcus lanatus L.
Hydrophyllum appendiculatum Michx.
Impatiens pallida Nutt.
Jeffersonia diphylla (L.) Pers.
Liatris cylindracea Michx.
Lilium superbum Michx.
Lindera benzoin (L.) Blume
Luzula multiflora (Retz.) Lejeune
Lycopodium lucidulum var. occidentale (Clute) L. R. Wilson
Magnolia acuminata L.
Muhlenbergia racemosa (Michx.) BSP.

Panicum barbulatum Michx.
Panicum implicatum Scribn.
Paspalum laeve var. pilosum Scrib.
Phlox divaricata L.
Poa alsodes A. Gray

Polygala verticillata L.

NORTHEASTERN UNITED STATES *Continued*

Polygonum exsertum Small
Psoralea onobrychis Nutt.
Pycnanthemum incanum (L.) Michx.
Ranunculus septentrionalis Poir.

Rhus typhina L.
Ribes cynosbati L.
Ribes missouriensis Nutt.
Rubus enslenii Tratt.
Rubus frondosus Bigel.

Rudbeckia fulgida Ait.
Rudbeckia hirta L.
Rudbeckia serotina Nutt. var. lanceolata (Fisch.) Fern. & Schub.
Salix fragilis L.

Saxifraga forbesii Vasey

Saxifraga pennsylvanica L.
Silphium terebinthinaceum Jacq.
Solidago canadensis L.
Solidago canadensis L. var. hargeri Fern.
Solidago graminifolia (L.) Salisb.

Solidago patula Muhl.
Tradescantia virginiana L.
Trillium flexipes Raf.
Valeriana pauciflora Michx.
Viburnum lentago L.

Viburnum recognitum Fern.
Vicia cracca L.
Viola sororia Willd.

SOUTHWESTERN UNITED STATES

Amorpha canescens Pursh
Asclepias viridiflora Raf. var. linearis (Gray) Fern.
Aster anomalus Engelm.
Aster turbinellus Lindl.
Chloris verticillata Nutt.

Clematis pitcheri T. & G.
Crataegus engelmannii Sarg.
Desmodium illinoense Gray
Eleocharis macrostachya Britt.
Equisetum hyemale L.

Equisetum laevigatum A. Br.
Eragrostis reptans (Michx.) Nees
Gaura biennis L. var. pitcheri Pickering
Jussiaea repens L. var. glabres-

cens Ktze.
Leptochloa attenuata (Nutt.) Steud.

Mentzelia oligosperma Nutt.
Monarda bradburiana Beck.
Panicum malacophyllum Nash
Paspalum stramineum Nash
Penstemon arkansanus Pennell
Prunus mexicana Sarg.

Ranunculus harveyi (Gray) Britt.
Sagittaria brevirostra Mack. & Bush
Scutellaria ovata Hill.
Solidago radula Nutt.
Talinum parviflorum Nutt.

Vernonia baldwinii Torr.

Acalypha ostryaefolia Riddell
Aesculus discolor Pursh
Agrostis elliottiana Schult.
Andropogon virginicus L.
Asclepiodora viridis (Walt.)
  Gray

Boltonia diffusa Ell. var. interior
  Fern. & Grisc.
Bumelia lanuginosa (Michx.)
  Pers.
Carex flaccosperma Dewey
Carex physorhyncha Liebm.
Carya texana Buckl.

Celtis laevigata Willd.
Cephalanthus occidentalis L.
  var. pubescens Raf.
Cheilanthes lanosa (Michx.)
  D.C. Eat.
Clitoria mariana L.
Cocculus carolinus (L.) DC.

Crotonopsis elliptica Willd.
Cynoglossum virginianum L.
Daucus pusillus Michx.
Draba brachycarpa Nutt.
Draba cuneifolia Nutt.

Fimbristylis baldwiniana
  (Schultes) Torr.

Galium virgatum Nutt.
Gerardia fasciculata Ell.
Helenium tenuifolium Nutt.
Heuchera americana L. var. interior Rosend., Butt. & Lak.

Hydrolea affinis A. Gray.
Hypericum lobocarpum Gattinger
Iresine rhizomatosa Standl.
Iris fulva Ker

Juncus diffusissimus Buckl.
Jussiaea leptocarpa Nutt.
Krigia dandelion (L.) Nutt.
Leptochloa filiformis (Lam.)
  Beauv.
Ludwigia glandulosa Walt.
Myosotis macrosperma Engelm.

Nyssa aquatica L.
Oenothera linifolia Nutt.
Panicum xalapense HBK.
Phaseolus polystachyus (L.)
  BSP.

Philadelphus pubescens Loisel.
Planera aquatica (Walt.) J. F.
  Gmel.
Polygonum opelousanum Riddell.
Ptilimnium nuttallii (DC) Britt.

Rhynchospora corniculata
  (Lam.) Gray var. interior
  Fern.
Ruellia pedunculata Torr.
Sedum pulchellum Michx.
Serinia oppositifolia (Raf.)
  Ktze.
Sibara virginica (L.) Rollins

Smilax bona-nox L.
Smilax bona-nox L. var. hederaefolia (Beyrich) Fern.
Smilax tamnoides L. var. hispida
  (Muhl.)
Solidago drummondii T. & G.
Solidago strigosa Small

Spermacoce glabra Michx.
Talinum calycinum Engelm.
Tragia cordata Michx.
Triodanis biflora (R. & P.)
  Greene
Viburnum rufidulum Raf.

Acer drummondii H. & A.
Agave virginica L.
Ampelopsis arborea (L.) Koehne
Ampelopsis cordata Michx.
Amsonia tabernaemontana Walt.

Andropogon elliottii Chapm.
Aristida oligantha Michx.
Aristolochia serpentaria L.
Aristolochia tomentosa Sims

Arundinaria gigantea (Walt.) Chapm.
Asclepias perennis Walt.
Asclepias variegata L.
Asclepias viridiflora Raf.
Ascyrum hypericoides L. var. multicaule (Michx.) Fern.

Cabomba caroliniana Gray
Calycocarpum lyonii (Pursh) Gray
Cardiospermum halicacabum L.
Carex decomposita Muhl.
Carex oxylepis Torr. & Hook.
Carex striatula Michx.

Carex styloflexa Buckl.
Carpinus caroliniana Walt.
Carya aquatica (Michx. f.) Nutt.
Cladrastis lutea (Michx. f.) K. Koch

Commelina diffusa Burm. f.
Corallorhiza wisteriana Conrad
Coreopsis pubescens Ell.
Cornus foemina Mill.
Crataegus collina Chapm.

Cyperus filicinus Vahl
Cyperus pseudovegetus Steud.
Desmodium viridiflorum (L.) DC.

Diodia virginiana L.

Dioscorea quaternata (Walt.) G. F. Gmel.
Erianthus alopecuroides (L.) Ell.
Gaura filipes Spach.
Halesia carolina L.

Heteranthera reniformis R. & P.
Heuchera parviflora Bartl. var. rugelii (Schttlw.) Rosend., Butt. & Lak.
Hexalectris spicata (Walt.) Barnh.
Hottonia inflata Ell.
Hydrophyllum macrophyllum Nutt.

Hymenocallis occidentalis (Le Conte) Kunth
Hypericum denticulatum Walt. var. recognitum Fern. & Schub.
Hypericum drummondii (Grev. & Hook.) T. & G.
Ilex decidua Walt.
Jussiaea decurrens (Walt.) DC.

Kuhnia eupatorioides L.
Liatris squarrosa (L.) Willd.
Linum striatum Walt.
Liquidambar styraciflua L.
Luzula echinata (Small) F. J. Herm.

Muhlenbergia capillaris (Lam.) Trin.
Muhlenbergia glabriflora Scribn.
Obolaria virginica L.
Panicum agrostoides Spreng. var. condensum (Nash) Fern.
Panicum anceps Michx.

SOUTHEASTERN UNITED STATES Continued

Panicum sphaerocarpon Ell.
Panicum yadkinense Ashe.
Paspalum dissectum L.
Paspalum laeve Michx.
Passiflora incarnata L.

Passiflora lutea L. var. glabriflora Fern.
Penstemon calycosus Small
Phacelia bipinnatifida Michx.
Phlox glaberrima L.
Phoradendron flavescens (Pursh) Nutt.

Pinus echinata Mill.
Polygonum setaceum Bald.
Polypodium polypodioides (L.) Watt.
Ptilimnium costatum (Ell.) Raf.
Pycnanthemum pycnanthemoides (Leavenw.) Fern.

Quercus falcata Michx.
Quercus falcata Michx. var. pagodaefolia Ell.
Quercus lyrata Walt.
Quercus michauxii Nutt.
Quercus phellos L.

Quercus shumardii Buckl.

Ranunculus pusillus Poir.
Sedum telephioides Michx.
Senecio glabellus Poir.
Smilax glauca Walt.

Smilax pulverulenta Michx.
Solidago buckleyi (Muhl.) T. & G.
Spiranthes cernua (L.) Rich.
Spiranthes ovalis Lindl.
Spiranthes tuberosa Raf. var. grayi (Ames) Fern.

Styrax americana Lam.
Synandra hispidula (Michx.) Baill.
Taxodium distichum (L.) Rich.
Tilia heterophylla Vent.

Tipularia discolor (Pursh) Nutt.
Tradescantia subaspera Ker
Triadenum walteri (Gmel.) Gl.
Trillium cuneatum Raf.
Ulmus alata Michx.

Urtica chamaedryoides Pursh
Vaccinium arboreum Marsh.
Vulpia octoflora (Walt.) Rydb.
Wolffiella floridana (J.D.Sm.) C. H. Thompson

PART II Vegetation

Nature's Design

One of the greatest attractions animals have to man is their behavior. They are dynamic and interesting to observe. Plants, by contrast, are noticed for their unusual shapes, colors, or the mass effect of their great number. To those who look carefully and often there is dynamic behavior also revealed among plants, not the individuals so much as the integrated assemblies we call vegetation.

Origin and Development of Vegetation

From pre-existing plant life a barren area may become populated. This early assembly, indeed the first occupants, constitutes vegetation. The coming together of individual plants on an area is ruled quite by chance, but their later performance is governed rigidly by the conditions imposed by their environment. The complete interaction of environmental influences upon all organisms exerts a selective and controlling effect. As organisms are thus exposed, some will do well where they are, others less well, some seem not to prosper at all, but remain because their lower range of tolerance to conditions is being met. A time is reached in these early developments of vegetation when the habitat, after successful migration of plants, becomes thickly populated with individuals. This close aggregation of plants results in competition. These competitive situations are severe, particularly when all residents are of the same species. Under these conditions all plants are making the same demands for light energy supplies, water, and nutrients.

As each plant asserts its demands, the habitat may become changed. It may become more shaded, more heavily mulched, more moist as in the case of individuals starting life under dry conditions. In the case of plants which begin their life under wet conditions, these habitat changes will trend toward drying. As a consequence of these plant reactions, the conditions of the many environmental situations may become changed sufficiently to cause some shiftings in the populations.

Some species are replaced by others more suited to new conditions so that a new community is developed. This replacement of one community of plants by another is a phenomenon called plant succession. Several such replacements may occur until finally the occupant plants produce ameliorating influences which become residual. There is some shifting in the community. There is still some coming and going, but always the kinds of plants remain in their greater or lesser numbers. "Everything flows and everything flees; nothing abides" (Heraclitus). Nature is still restive and only dynamic stability is reached.

In any given place a definite sequence or series of plant communities will occur in the march of vegetation to its highest development. The cause for this vegetational change in an area includes a host of factors such as changes in habitat, chance migration, competition, release of toxic and inhibiting substances, the life cycle of some species, etc. Thus as new plants come into an area they manifest further change by their reactions upon one another and upon the habitat. This reaction may be interpreted as the driving force in the process of succession (Weaver and Clements, 1938). It is axiomatic that each organism renders its habitat unfit for its continued existence and more fit for another organism to follow. This is commonly and convincingly seen in the inhibition of the growth of bacteria when their massed toxic metabolic wastes prevent an unlimited proliferation. Man would seem to be no exception despite his intelligence to do otherwise. How familiar the country roads strewn with rubbish and dump heaps, the once crystal streams now colored with pollution, or cities themselves with slums as testimony of decay and a final stage in a cycle. Since a series of steps or stages is involved in ecological succession of an area, the total organized development has been termed a sere.

Seres are of two kinds. The primary sere is one which involves a series of plant community replacements in an area where vegetation has not existed previously. The secondary sere or subsere is one where vegetation has existed, but has been removed, although the effects of previous occupation are still present in the area. Commonest primary successions are those of open water, such as ponds and lakes, and bare rock outcrops. The development of vegetation in a water situation is termed a hydrosere, and the development of land plants on an arid rock surface is a xerosere.

Observation of an open body of water reveals an early development of submerged plants. Later, perhaps a matter of several or many years, there will develop floating forms. Some of these float free on the water surface while others have floating leaves but some attachment to the bottom in shallow water. As these plants either die or still the currents of water draining into the lake, the bottom is built up. Accumulations of organic materials and inwashed sediments cause a gentle gradient from deeper water to the shores. Into these newly created shallows near the shore come plants such as cattails (*Typha* spp.), bulrushes (*Scirpus* spp.), arrowheads (*Sagittaria* spp.), etc., which grow with their roots in the mud and their leaves above the water. The effects of these "amphibious" plants are spectacular. They dry up the water through their vigorous transpiration and add literally tons of organic residues as well as some mineral matter to the bottom each season. A wet meadow condition follows and is inhabited by spike rushes, sedges, grasses of various kinds, wild iris, etc. Later the environment becomes changed still further and is suitable for growth of shrubs, then scattered trees, and ultimately a forest. These stages often exhibit a well-marked zonation (Fig. 16). The early stages are always similar, but the end-point of vegetational development is respondent to climatic conditions of an area. Thus a hydrosere west of the Mississippi River in the midcontinental region which is climatically drier will terminate in grassland.

Climax

These dynamically stabilized communities, highest vegetational types possible under particular sets of climatic conditions, are known as climatic climaxes. In them, certain plants by virtue of their greater size, perennial nature, vigor of growth, reproductive capacity, and superior numbers will have achieved a position of dominance. There are, of course, many other plants in the vegetational fabric which have adjusted themselves to a life of subordination or to a role of being subdominant. Collectively the representatives of this climax state live in harmony, and barring any profound changes in either climate or substratum, the climax community tends to be self-maintained. In its collective harmony and full use of environmental resources this end-point of community development has been termed a "closed community." This means that the community is not open to foreign or invading plants because its energy supplies are being utilized so completely. Thus roadside weeds may occur on one side of a fence but not in an adjacent climax prairie (Weaver & Clements, 1938).

As in all natural phenomena, the process of succession is continuous. If we could pass in review the work of succession in the manner of viewing moving pictures—each frame of film being equivalent to

16. Diagram of the zonation of a hydrosere showing, from left to right, the stages of submerged, floating-leaf, reed-swamp (amphibious), wet-meadow, shrub, and tree vegetation.

work of a year or several years—it would take but a few hundred frames to tell the story—a matter of minutes on film, but a few hundred years actually. These time-lapse intervals may be observed in four or five ponds of different age in an area, or in several rock ledges of related topographic development in the case of a xerosere.

Generalized Succession of Hydrosere in Southern Illinois

I. Submerged Stage
 Carolina Water-shield
  (*Cabomba caroliniana*)
 Hornwort
  (*Ceratophyllum demersum*)
 Water-weed
  (*Elodea canadensis*)
II. Floating Leaf Stage
 a. Free Floaters
 Water Fern
  (*Azolla caroliniana*)
 Duckweed
  (*Spirodela polyrhiza*)
  (*Lemna trisulca*)
  (*Lemna valdiviana*)
  (*Lemna perpusilla*)
  (*Lemna minor*)
  (*Wolffia columbiana*)
  (*Wolffia papulifera*)
  (*Wolffiella floridana*)
 b. Attached Floaters
 Yellow Pond Lily
  (*Nuphar advena*)
 White Water Lily
  (*Nymphaea odorata*)
 Water-shield
  (*Brasenia schreberi*)
III. Amphibious Stage
 Cattail
  (*Typha latifolia*)
 Bur-reed
  (*Sparganium eurycarpum*)
 Water Lotus
  (*Nelumbo lutea*)
 Bulrush
  (*Scirpus validus*)

IV. Wet Meadow Stage
     Iris
        (*Iris brevicaulis*)
     Spike rush
        (*Eleocharis obtusa* var. *engelmannii*)
     Sedge
        (*Carex vulpinoidea*)
V. Shrub Stage
     Buttonbush
        (*Cephalanthus occidentalis*)
VI. Tree Stage
     Cottonwood
        (*Populus deltoides*)
     Black Willow
        (*Salix nigra*)
     Sycamore
        (*Platanus occidentalis*)

Attention is called by Buell (1949) to an outstanding account of primary succession in 1792 by Willdenow, a great German botanist. The primary succession in this case is that of a xerosere. "Naked rocky places, on which nothing can grow, are, by the winds, covered with the seeds of lichens, that by means of the accustomed showers in Harvest and Spring are induced to germinate. Here they grow, and the rock is spotted with their colored frond. In time the winds and weather deposit small dust in the rough interstices of rock, and even the decaying lichens leave a thin scurf. On this meagre soil the seeds of mosses are accidentally driven, where they germinate. They grow and produce a pleasant green tuft, which in time is fit for the reception of the smaller plants. By the rotting of the mosses and small plants, there arises a thin layer of earth, that in the course of time increases, and then becomes fit for the growth of various shrubs and trees, till at last, after many years, where formerly there was nothing but naked rocks, the eye of the traveler is gratified with the sight of extensive Nature. Gradual, great, and constantly conducive to general good are her operations. Mosses and Lichens improve in a similar manner the dry and barren sands. The plants that grow naturally in such soils have almost all creeping and extensively penetrating roots; or they are succulent, and draw moisture from the atmosphere. By means of these plants the sandy soil is made fit for the reception of mosses and lichens, and afterwards change into good and fertile earth." (Translation by Murray F. Buell, *Ecology*, vol. 20, no. 1, 1949.)

The foregoing describes the classical concept of a primary succession on rock or of a xerosere. It has been found, however, that the xerosere succession is not always initiated by lichens and mosses, but may begin with annual and/or perennial herbs (Winterringer and Vestal, 1956). On granite rocks of North Carolina lichens did not play any important part in the development of later stages. Lichens are pioneer species, but they were found to be rarely, if ever, followed by other forms. Anywhere on the rock surface there seemed to be black moss (*Grimmia*), reindeer lichen (*Cladonia*), resurrection plant (*Selaginella*), haircap moss (*Polytrichum*), bluestem or broomsedge (*Andropogon*), and conifer development. Dry depressions usually supported flower-of-an-hour and sedges (*Fimbristylis* and *Stenophyllus*) (Oosting and Anderson, 1939).

Generalized Succession of a Xerosere in Southern Illinois

I. Lichen Stage
    Crustose lichen
    Foliose lichen
    Fruticose lichen
II. Moss Stage
    Black Moss (*Grimmia*)
    Haircap Moss (*Polytrichum*)
    Twisted Moss (*Tortula*)
III. Herbaceous Stage
    1. Annuals
        Wire grass (*Aristida*)
        Sundrops (*Oenothera linifolia*)
        St. John's-wort (*Hypericum gentianoides*)
    2. Perennials
        Sedum (*Sedum pulchellum*)
        Agave (*Agave virginica*)
        Cactus (*Opuntia rafinesquii*)
IV. Shrub Stage
    Farkleberry (*Vaccinium arboreum*)
    Catbriar (*Smilax*)
    Buckbrush (*Symphoricarpos orbiculatus*)
V. Tree Stage
    Red Cedar (*Juniperus virginiana*)
    Post Oak (*Quercus stellata*)
    Black Jack Oak (*Quercus marilandica*)

Sere Convergence

Owing to diversity of seral conditions in primary successions over a regional area, it is to be expected that there will be similarly an equal amount of variation in early seral communities. These pioneer communities will, under the reign of a more or less uniform regional climate, show a successional development characterized by a considerable uniformity of medial or late stages. The development of vegetation in diverse habitats to a more uniform condition and similarity of plant communities is known as convergence (Clements and Shelford, 1939).

Thus, in succession of a hydrosere, through plant reactions, a development from excess of water to a moderate amount of moisture is shown. Other habitat factors, too, undergo such change from extreme to moderate. By the same token, succession of a xerosere shows the same tendency to medial conditions, but begins at the opposite end of the scale, that of extreme dryness. Such convergence results in climax communities of medial habitat character.

In southern Illinois such diverse upland habitats as sandstone ledges and clay hills converge to an upland climax of oak and hickory. Aquatic habitats (deep and shallow) and floodplains converge to a mesophytic ravine-type climax dominated by beech and maple. Given sufficient time in the erosion cycle for leveling of uplands and genetic stability of plants during this period, the climax would probably be some compromise between present upland and lowland climax communities with dominance being achieved by the most shade-tolerant types. With great but immeasurable length of the time interval and susceptability of species to genetic change with changed habitats and over long periods of time, the climax is presently considered by many (Whitaker, 1953) to be a mosaic or pattern of climaxes along environmental gradients.

In southern Illinois, not much less than a hundred years ago, there prevailed many acres of rich forest. As land was cleared and cultivated, soil was changed profoundly. Changes in organic matter, nutrients, and structure were caused by leaching, runoff, erosion, and the cropping systems.

As these changes were manifested over a period of years, the land became unproductive and was often abandoned in favor of homestead land to the West. After lands were abandoned, the plants restored to them were not immediately those which had originally occupied the land. During the first few years of revegetation, the plants are conspicuously weedy species, mostly annuals. Later, perhaps after a half dozen years, perennial herbs, grasses, and some brambles may appear. Still later, shrubs make their appearance and then the vanguard of tree species appears. Earliest appearing trees in our area are often sassafras and persimmon or red cedar whose seeds may be spread by birds which frequent the open aspect of bramble fields. Little variety of undergrowth exists under trees until stands thicken, compositions change, and reactions continue to the point where rich leaf mold restores conditions of the ground layer in the original forest. Only then may we compare the vegetation with that which was originally there.

Where former vegetation has been removed by cultivation and through drought, fire, overgrazing, etc., and through subsequent abandonment or change in land-use policy the land is repopulated by successive plant communities. The process is known as secondary succession, and is in operation everywhere and in all types of vegetation.

All students of secondary succession are not in agreement on length of time needed for a climax to be restored, but are agreed, however, that it takes a long time. Studies in grassland areas indicate from twenty to sixty years are needed (Shantz, 1917).

In forest areas, most studies indicate a subclimax tree stage may be reached in thirty to sixty years and that several generations of competitive selection must pass before a reasonable climax structure and composition are attained. Thus it is suggested, because of longevity of species in each generation, that it is realistic to view the end-point in forest succession in terms of several hundred years. An outline of secondary succession is presented to reveal the timetable and organization of succession as it might be in southern Illinois.

The Probable Secondary Succession in Southern Illinois

1. Annual Weed Stage (usually 1 to 3 or 5 years)
    Annual Ragweed (*Ambrosia artemisiifolia*)
    Biennial Primrose (*Oenothera biennis*)
    Annual Panic Grass (*Panicum dichotomiflorum*)
    Brome Grass (*Bromus tectorum*)

II. Perennial Weed Stage (usually 3 to 10 years)
   Frost Aster (*Aster ericoides*)
   Broomsedge (*Andropogon virginicus*)
   Goldenrod (*Solidago nemoralis*)
   Bluegrass (*Poa pratensis*)
III. Shrub Stage or Shrub and Bramble Stage (usually 7 to 15 years)
   Poison Ivy (*Rhus radicans*)
   Sumac (*Rhus glabra, R. copallina*)
   Dewberry (*Rubus flagellaris*)
   Blackberry (*Rubus allegheniensis*)
IV. Pioneer Tree Stage (usually 10 to 25 years or before)
   Persimmon (*Diospyros virginiana*)
   Sassafras (*Sassafras albidum*)
   Black Locust (*Robinia pseudoacacia*)
   Pignut Hickory (*Carya glabra*)
V. Later Tree Stage (usually after 30 years)
   Red Oak (*Quercus rubra*)
   White Oak (*Quercus alba*)
   Black Oak (*Quercus velutina*)

Succession has been one of the most fruitful concepts of plant ecology. Direction of succession by man's alteration of natural conditions is the very essence of such pursuits as agriculture, forestry, range management, and wildlife management. By allowing overgrazing, man has permitted degeneration of vegetation which is also a process of recognizable stages (Voigt and Weaver, 1951). By adjusting grazing pressure through control of animal numbers or otherwise establishing protection to plants, a forward direction of vegetational development or succession may be re-established. By permitting flooding in northern states, bogs are maintained for commercial cranberry production. Fire has been used as a tool in establishing a pine sub-climax along the Atlantic Coastal Plain where unburned areas develop oak and hickory communities. Here, thin-barked oaks are killed by fire, whereas pines having thicker bark are favored. Nearly pure stands of pine result from periodic controlled burning. Wildlife habitat may be variously modified by burning, cutting, mowing or not mowing, grazing or not grazing, planting of certain food plants, etc. In all aspects of vegetational management, it is imperative to know plant compositions of climax and all stages of development and to understand relationships of disturbance or causal factors (animals, fire, flooding, land use policy, etc.) to each. Thus management of plant communities is based upon our knowledge of behavior of plant species under varied conditions and upon a knowledge of community compositions and their successional behavior under various natural and artificial influences.

In America, it was H. C. Cowles (1898) who first demonstrated plant succession in his study "Ecological Relation of Vegetation of the Sand Dunes of Lake Michigan." Cowles recognized certain aggregations of plants and found them to correlate with topographic development. Though not the earliest use of the successional story, Cowles' work demonstrated the phenomenon most clearly and did much toward establishing the succession concept as one of the most useful principles related to vegetational study.

Coincident with Cowles' work, Roscoe Pound and F. E. Clements (1898) published their "Phytogeography of Nebraska." The employment of the quadrat as an instrument for detailed qualitative and quantitative measure of vegetation was an important contribution of this work. In knowing more or less exact compositions as of a given time, it was possible to follow succession in greater detail.

Structure of Vegetation

Original use of the phrase that one "could not see the forest for the trees" was undoubtedly facetious, but as applied to vegetation, was used to connote complexity or disorganized aspect of woodland. Any vegetation, whether woodland, grassland, or other, and of whatever stage of development or age, will be found to possess a characteristic structure. This structure may be recognized most easily in an older bit of vegetation where some degree of stability has been reached. To discover the underlying design, J. E. Weaver has stated that it is only necessary for a student or observer to follow two principles: "look carefully and look often."

Over most of the eastern part of the United States occurs a vast area of forest vegetation. Climate prevailing over this area is favorable to growth of trees, though marked differences in degree of favorableness to this growth occurs from place to place. These climatic divisions result in several kinds of forests, each recognizably different.

Climatic changes from north to south are geared mainly to the factor of temperature, and climatic changes from east to west are geared more to a declining amount of precipitation. Temperature and moisture are just two factors of many operating in multiplicity.

Vegetation responds to these climatic conditions strikingly, sometimes less so, when change is more subtle. Experience largely determines ability of the observer to pick out vegetation types, but it is not difficult to recognize the regional character of vegetation.

If one were to travel southward from Maine to Florida, northern forests of hemlock (*Tsuga canadensis*), white pine (*Pinus strobus*), and northern hardwoods of beech and sugar maple (*Acer saccharum*), birch (*Betula* spp.), basswood (*Tilia glabra*), and others would be replaced in New Jersey by the growth of pines of the Coastal Plain. These pines are a subclimax which extends to Florida, around the Gulf, and into the Mississippi embayment. The main species of this type are pitch pine (*Pinus rigida*) in the North, longleaf pine (*P. palustris*), loblolly pine (*P. taeda*), and shortleaf pine (*Pinus echinata*). Poor drainage in the southeast, along with other factors, soon presents swamp communities of bald cypress and gum and, in South Carolina and northern and central Florida, the cabbage palmetto (*Sabal palmetto*).

A journey from the hemlock, white pine, and northern hardwood forests west of the Coastal Plain would reveal several climatic areas by presence of other recognizable vegetation types. In the northern part of the vast deciduous forest, extending south to northern parts of the central eastern states, is a beech and maple region. West of this is a small forest region of maple and basswood. It occupies unglaciated territory and extends to south-central Minnesota. Southward of this and back eastward to the Coastal Plain and south of the beech-maple type is the largest deciduous forest type—the oak and hickory forests. This type crosses the Mississippi River southward into Missouri, Arkansas, Oklahoma, and Texas. The parent of all deciduous forest types, the mixed mesophytic, is centered in eastern Kentucky and Tennessee (Braun, 1950). It is surrounded by the oak-hickory type.

The forest is a highly constituted community. The organization and role played by dominant species, the close interaction between them, their subdominants, and animal components of the community under stable and unstable conditions are fascinating study. It is through the unity of dominant species that communities are recognizable, though it is sometimes observable that herbaceous constituents may also contribute to this distinction (Potzger and Friesner, 1940; Daubenmire, 1952).

Climatically, southern Illinois is suited to growth of forests dominated by oaks and hickories. These species, though not the only ones present, are dominant on drier, average topographic sites of upland. On low ground and in protected ravines, the dominants are perhaps beech, maple, sourgum, sweetgum, and others. Between are various combinations of these and other dominant species ranging from one, two, or three species on nearly equal terms to perhaps a dozen or more. Many different communities may thus be seen to make up the larger forest unit. Each such community is delimited by a complex of environmental factors operating in that particular area.

The main features of the forest vegetation of southern Illinois are dominance of trees and shrubs, scarcity of grasses except in openings or on some wet or disturbed areas, and wealth of spring flowers. All plants of the forest have reached an adjustment to the seasons, and in spring some species flower before leaves unfold. Summer activity of food-making and food-storage gives way to gorgeous color and fruition in fall. Four seasonal aspects are usually recognized. The pre-spring aspect begins in February and is followed by the spring aspect in about April, the summer aspect in June, and the fall aspect in late July or August. Earliest appearing plants spring up from edges of the forest and from situations in woods receiving greatest warmth from sunlight.

Habitats of all kinds exist in these forests. They range from open water to dry sandstone blufftops; from high elevations and brightly lighted areas to shaded undersides of cliffs and caves which may be dry or continually moist with seepage. There is rarely a shortage of living space in the forest; rather, there may be a shortage of light, water, or nutrients in the soil. In response to certain supplies of these habitat factors, vegetation has made another structural adjustment. Different levels of the soil are occupied by roots and different statures may be attained by plants above ground. This kind of adjustment is known as layering (Fig. 17). Usually three layers may be observed both above and below ground. The upper or canopy layer of the forest is occupied by dominants. Smaller trees and shrubs just beneath them constitute a mid-layer. The forest floor, occupied by the subdominant herbs, is known as the understory. Layering of dogwood and redbud in spring is a scene of unforgettable beauty. Similarly, spring woodland groupings of bloodroot, trilliums, and others are pleasingly beautiful.

17. [left] Layering in a forest west of Harrisburg in Saline County is obvious with flowering of redbud. [right] A back road south of Hickory Ridge in Jackson County presents numerous hills where vegetation alternates according to north-south slope aspect.

Slope aspect often manifests a structural adjustment in vegetation known as alternation (Fig. 17). Woodland on north slopes alternates with grassland on south slopes on bluffs along the Mississippi River in southern Illinois. Each of these strikingly different plant communities is a reflection of the contrasting slope environments (Weaver and Clements, 1938).

SUMMARY: *How Vegetation Shows Structure*

1 By attaining different life forms
2 By segregation into different plant communities
3 Through dominants
4 Through subdominants
5 By layering
6 Through adjustment to seasons (aspects)
7 Through zonation
8 Through alternation

Energy Relations

With greater exploration of energy relations in living organisms and their processes came a new viewpoint of plant or animal study known as physiology. This viewpoint was first exploited in explaining distribution of plants by A. F. W. Schimper (1903) in his book *Plant*

Geography Upon A Physiological Basis. Importance of the physiological point of view to a study of plant communities is more easily comprehended when the community is made analogous to an organism. The idea that a living plant community is like an organism, is an organism, or is in the same class of phenomena as an organism was set forth by Clements (1916). Though we understand a great deal about chemical and physical aspects of protoplasm, it is energy which provides the vital spark in manifesting behavior attributed to living things. Behavior is a product of form and function and therefore at the root of organic integration is the problem of energy (Sears, 1949).

The flow of energy through the plant community with its animals, indeed the operation of the whole of nature, has been envisioned as a system of debits and credits and lasting balance is achieved in the collective state of harmony called climax. Virgin forests or climax prairie and others with their wonderful organization, their great production of organic materials, are unmistakable testimony of energy use. A certain amount of energy has been used and an equal potential is held.

The association of richness and beauty is ages old. The less the energy level of the community the less likely we are to view it as something beautiful. It is the comparison of primeval vegetation with barren waste, where energy, instead of being abundantly stored, is re-radiated and lost. It is the contrast of luxuriant cover and production against erosion and loss.

As surely as heat energy flows from a region of intensity to one of a minimum, we may be sure that energy follows the same pattern in our economy. Energy richness in the economy is reflected in wealth. The many levels of the social order in our civilization is a reflection of the proportion and arrangement of wealth. The flow of energy turns the wheels of industry and moves the goods of trade. Aldo Leopold (1949), that great philosopher and spokesman for the conservation effort, has remarked to the effect that economics is basically aesthetics.

Tansley (1935) coined the term ecosystem to refer to complex interactions between organisms and physical factors in an area. Another way of stating this idea is that the ecosystem is the communities of organisms plus their habitats. Considered as such, the largest ecosystem with which we are familiar is earth itself. When a number of organisms is present in an area, certain interactions result. Primarily, green plants synthesize carbohydrates from sunlight, carbon dioxide, and water. Herbivorous animals feed upon them and carnivores in turn upon the herbivores. Metabolism of living organisms and death and decay of plant and animal tissues release materials to the soil to be used again. Thus many systems may be seen to operate in a cyclic manner, all being dependent in their operation upon the sun as a source of energy. The ecosystem concept emphasizes the flow of energy as related to communities in time and space.

The ecosystem concept which emphasizes inter-relatedness of the whole of nature has been expressed also by the suitable designations "Web of Life," "Balance of Nature," or the concept of a "Biotic Pyramid."

In the latter concept, energy absorbed by the green plants from the sun flows through the biota beginning with the soil as the broad base of the pyramid. Plants rest upon and in the soil, and are nurtured by it. Insects and other herbivores make up the next layer and so on to the carnivore layer at the top. The pyramid so described is based upon a reduction in numerical abundance of organisms in ascending order.

It is impossible for any organism to exist entirely unto itself. The pyramid of numbers emphasizes a broad linkage of different levels of organisms based upon their sizes and kinds. Many more intricate interlocking relationships exist within the levels themselves. Thus a complicated network exists which makes the phrase "live alone and like it" completely untenable.

Nature's guiding hand has touched all living forms for untold ages. The reactions made by each kind of organism upon others over a long period of time have sorted and selected those which can endure one another's company or indeed claim benefit from it. These reactions resulting from the infinite number of organismal associations are the forces which shape the pyramid.

It has been noted that rocks are corroded and softened by laboring lichens, that mosses capitalize upon this effort and occupy the softened rock areas and act in sponge-like manner to keep the streams flowing. Smaller plants are encouraged and entrap the sunlight and build the soil. Grasses protect the earth like a garment and forests temper the influence of the elements giving protective shelter to a multitude of living things.

Heraclitus, an obscure Greek philosopher writing about 500 B.C., noted these restive actions of nature and described the interconnective relations and flow of life as follows: "The rain falls; the springs are fed; the streams are filled, and flow to the sea; the mist rises from the deep and clouds are formed, which again break on the mountainside. The plant captures air, water, and salts, and with sun's aid, builds them up by vital alcheming into the bread of life, incorporating this into itself. The animal eats the plant and a new incarnation begins. All flesh is grass. The animal becomes part of another animal and the reincarnation continues. The silver cord of the bundle of life is loosed and earth returns to earth." The microbes of decay break down the dead and there is a return to air, water, and salts. We may be sure that nothing is ever really lost. We are sure that all things flow.

Classification of Vegetation

Just as plant taxonomists had to create ways in which to classify thousands of plants in the world, so the ecologist has had to devise means of classifying ecologically related groups of vegetation. Various schemes have been offered; the aim of each has been to provide a sound framework so that conditions of vegetation found in nature may be portrayed as realistically as possible.

Environment is the cause and the community is the effect. Wherever environment is made up of distinctive elements of unusual strength or intensity, the vegetation itself is distinctly expressed. Nearly every traveler has been aware of difference in landscape as he departs one region for another. In analysis of scenery it is recognized as being pre-eminently land forms which provide the framework, but it is the fabric of vegetation which gives land forms their special scenic character. The great geographer and "Universal Man," von Humboldt (1769–1859), in his travels observed many times the apparent relationships of climate to physiognomy or characteristic appearance of vegetation. His writings were later to be used as a basis for ecological classifications as, for example, rain forest, deciduous forest, prairie, desert, etc.

Humboldt's writings gave a strong and zonal emphasis as shown in his words—"Unlike in design and weave is the carpet which the plant world in the abundance of its flowers has spread over the naked crust of the earth, more densely woven where the sun ascends higher on the cloudless sky, looser toward the sluggish poles, where the early returning frost nips the undeveloped bud and snatches the barely matured fruit. Every zone is endowed with peculiar charm—

the tropical in the variety and grand development of its production, the northern in its meadows and fresh in the periodical revival of nature and the influence of the first breezes of spring. Besides having its own special advantages, every zone is marked by a peculiar character."

Correspondence of deciduous forest with one type of climate and grassland with another existed also in the mind of Grisebach who used the term community in describing certain vegetational units This community organization also was noted by keen-minded and observant Anton von Kerner in his travels through the Danube Basin. Kerner's work in many places appears unquestionably the basis for much of the modern ecological theme. (Conard, 1951).

The nature of the more prominent plants which enter into the plant community determines the physiognomy. More accurately, physiognomy is identified according to the dominant life-forms of plants. Life forms may be considered as the form and structure by which a plant meets the problems of existence in different environmental conditions.

In its simplest terms the vegetation of the earth may be described as being composed of three types. It is either some kind of forest, some kind of grassland, or some kind of desert. Many in-betweens of these basic types may be found from place to place depending upon the prevailing environmental conditions. In all the different kinds of forests the dominant plants are trees, in grasslands the dominants are grasses, and in deserts scanty herbaceous plants or widely spaced shrubs or both. Hence from the standpoint of dominant plants there are only three life forms: trees, shrubs, and herbs.

Remembering that morphological features are usually reflections of functions, we may thus expect a strong morphological similarity of individual plants even in widely separated areas if climate of the two areas is similar. The life forms of a region climatically similar to another could be compared in terms of numbers or percentages of life forms. The figures thus obtained for each life form class constitutes the biological spectrum (Raunkiaer, 1934). A comparison of plants of two areas, Spitsbergen, Norway, and St. Lawrence Isle, Alaska, shows nearly identical spectra, though the species are different. Another close comparison in biological spectra is in the plants of Death Valley, U. S. A., and Argentaria, Italy; both have a mediterranean type of climate (Sampson, 1952).

Raunkiaer's scheme, based upon adaptations of species to endure an unfavorable season, will be mentioned briefly here. Raunkiaer divided all plants into five groups. His Therophytes included those species completing their life cycle in one growing season and passing through the unfavorable season in the seed stage. Cryptophytes are those plants whose buds survive the unfavorable season in the ground. Perennial bunch-forming plants whose shoots die back to the ground at the beginning of the unfavorable season are classed as Hemicryptophytes. Chamaephytes include those species whose buds lie near the surface of the ground (from decumbent shoots or stolons). The fifth category, Phanerophytes, is composed of usually woody plants whose buds extend into the air and are not conspicuously protected.

Ecological structure includes all peculiarities in vegetation which are of ecological significance. Not only may physiognomy be included here, but also behavior of plants in relation to various conditions of light, temperature, humidity, water content of the soil, etc. Ecological structure is more inclusive than life-form in that it considers quantity and quality of numerous subordinate life-forms within the community. Various shrubby and herbaceous species would be included under ecological structure where they would be unaccountable in physiognomy.

In comparing areas which differ markedly from one another in either climate or physiography, one may easily observe the effect which geographic factors have on vegetation. The assemblage of species in the tundra is vastly different from that of tropical rain forest, the differences being due, for the most part, to the gross difference in climate of the two regions. Differences resultant from physiography are revealed in comparing vegetation of a ravine with that of a floodplain, etc.

Climate, physiography, and biota are the chief habitat factors which shape the environment. Let us consider climatic conditions first. Because of their variation from habitat to habitat, the factors of moisture, temperature, and light are of greatest importance. Availability of moisture influences not only rate of plant growth but also growth form. It also controls geographic distribution of plants to a certain extent. Species such as river birch and alder are seldom found away from stream or river banks because their seeds perish rapidly if they do not fall upon a very moist soil. Other species may have seeds which are capable of existing in a dormant state for a considerable period of time, germinating only when sufficient moisture is available. Thus Clements (1904) and others designated various vegetational areas throughout the United States—prairie, deciduous forest, desert, etc. These climaxes are due to climatic conditions of the area, primarily with respect to moisture availability.

Temperature not only affects each individual plant, but also vegetational communities. It has a marked influence on the establishment of a plant in an area, although the individual species within each are very different.

Another important factor in local distribution of plants is intensity of light since this varies widely in different macro-environments. Photoperiod or length of day is likewise important in distribution of plants. Thus plants which require long days in which to flower, and subsequently reproduce, do not persist in short-day regions. Conversely, plants adjusted to short day flowering are not found naturally occurring in long-day regions such as the higher latitudes.

Where climate is uniform, differences in vegetation are directly related to differences in topography and soil. Vegetation has developed during the course of physiographic advancement with the result that the various local habitats are arranged in a definite manner in relation to larger physiographic features of the region.

Principal physiographic habitat factors which affect form, structure, and behavior of the earth's surface are topographic and edaphic in nature. The effect which topography has on the habitat is expressed directly through slope and indirectly through modification of other environmental factors (Nichols, 1923).

Irregularities in topography such as bluffs, ridges, and depressions cause considerable differences in moisture, temperature, and light conditions. The environment of north-facing and south-facing slopes usually differs greatly enough to support distinct plant communities. South-facing slopes are drier, have higher temperatures, and receive more light than those facing north or even average sites in the area. In southern Illinois, plant communities on south-facing slopes are dominated by pignut hickory, red oak, and black oak. North-facing slopes usually are dominated by communities of tulip tree, sugar maple, and beech, while the understory is densely populated with a large number of herbaceous species.

Edaphic factors include physical structures and chemical composition of soil, or whatever medium plants grow on or in. Physical structure of the soil includes particle size (sand vs. clay vs. rock, etc.), while chemical composition is concerned with pH, mineral content, etc. Factors which bring about changes in physical structure and chemical composition of soil include erosion, deposition, and soil leaching.

A physiographic area in reality represents a series of habitats related by physiographic development. Topographic conditions seem to be of greatest ecological significance, but edaphic factors also exert a considerable influence in developing the habitat type. A physiographic classification of plant communities must therefore be concerned with both topographic and edaphic conditions. Such a system was proposed by Cowles (1901) who sought a "genetic and dynamic" classification. An attempt was made to group plant communities according to their relationship and evolution. In his "Physiographic Ecology of Chicago," Cowles devised the following classification headings:

A. The Inland Group
  1. The river series
     a. The ravine (clay and rock)
     b. The bluff (sandstone and limestone)
     c. The floodplain
  2. The pond-swamp-prairie series
     a. The pond
     b. The undrained swamp
     c. The prairie
  3. The upland series
     a. The rock hill
     b. The clay hill
B. The Coastal Group
  1. The lake bluff series
  2. The beach-dune-sandhill series
     a. The beach
     b. The embryonic or stationary beach dunes
     c. The active or wandering dunes
     d. The established dunes

It is nearly impossible to draw dividing lines between different physiographic areas. For example, scattered throughout rocky uplands along the Shawneetown Ridge of southern Illinois may be found small areas with swamp or ravine formations, these being due to minor variations in topography.

On the other hand, there are sometimes areas which are physiographically distinct that support similar vegetation. To understand this, we realize that one habitat factor may be compensated by the influence of another habitat factor. Thus the amount of soil moisture in one area may be great enough to offset physiographic differences between two areas so that the resultant vegetation is essentially alike.

Biotic habitat factors include those which are associated with the activity or effect of plants and animals, including man. Often man exerts such an influence on vegetational communities that his actions are sometimes classified in a separate category termed anthropeic factors (Nichols, 1923). Biotic factors have caused the development of certain types of vegetational communities. Thus areas which have been cleared for cultivation and then no longer are used for that purpose soon grow up into discrete plant units termed by various authors as waste ground communities, old field communities, etc. Distinct plant communities likewise develop after fires which have a natural or anthropeic origin.

To close out the discussion of classification of plant communities, two systems which have developed during the late 1940's and early 1950's will be reviewed. Curtis and McIntosh (1951) have presented their views under the title of the continuum, an expansion of the work of Gleason in 1926. Gleason thought that because individual organisms of a species are subjected to chance migrations and subsequent chance survival in their new areas, they will segregate from their kind and other associates in all manner of ways. He recognized the importance of this individual response and thought that no community reaches any definite composition. According to this idea, each individual is the "community." Curtis explains segregation of species along environmental gradients. These gradients provide a gradual interplay of forms so that it is impossible to delimit communities.

Community recognition is based upon composition and is largely subjective. If the composition consisted of only two or three leading species and if there were an abrupt change in composition, the community boundaries would be sharply drawn. In nature, however, species will generally not be so few and the lines are much less sharply drawn. Changes may be subtle and occur over a distance so that we may say change follows a gentle environmental gradient. In choosing plant communities from such a situation, it is apparent that boundaries are apt to be drawn subjectively and arbitrarily. There are many situations in nature, however, where recognizable communities exist even to those with relatively little experience in selecting them. The ease of recognition depends upon the experience of the individual and upon the degree of detail imposed by him in his attempt to separate the various groupings. Even in such a complex vegetation as grassland it is necessary to know only about a dozen dominant species to grasp community structures. The older community concept has enjoyed a certain amount of popularity and success in application to land management problems and therefore may be expected to live on. The existence and certainty of a continuum is well established and remains yet the closest approach to displaying all facts at once.

Daubenmire's work (1952) with forest vegetation of Idaho was stimulated to a large degree "from the absolute necessity for a critical delimitation of vegetation units in order to organize and relate studies of the causes of vegetation distribution." The system by Daubenmire was a composite of favorable features of several concepts.

Daubenmire considered the association as the basic unit of vegetation and applied this term only to climax communities. It is composed of unions (layers) which are the smallest structural units in the organization of vegetation. It delimits the community types as controlled by habitat differences, better than the use of trees alone. Each unit consists of one species or several species related in their ecological requirements. Thus his *Thuja-Tsuga/Oplopanax* association indicates a dominant union typified by *Thuja* plus *Tsuga*, combined with a subordinate union typified by *Oplopanax*. This classification presents a brief diagnostic description and depicts physiognomy to a certain extent.

Methods of Studying Vegetation

The term community is a very flexible bit of nomenclature in ecological usage. It is synonymous with type which may be used with greater ease most of the time. In either case the connotation is a stand of vegetation; it may be a very specific kind of stand or a mixture of several kinds of vegetation. The community may be large or it may be small. It may be of dominants, mixtures of dominants and subdominants, or of subdominants alone. Characteristics of larger communities of vegetation such as formations and associations have been well known to ecologists for a considerable time, but the existence of many smaller and often complex types within larger units raises the need for a measure of these smaller types so that they may be better recognized and understood.

In interpreting vegetational change or succession, it is necessary to have a fixed point of reference. This, for the ecologist, is usually a regional climax. To determine the degree of change necessitates methods of study which give details of growth, composition, and other qualities that constitute structure of vegetation as of a particular place and time. Such detailed knowledge is also needed to accurately judge cause and effect relationships.

Few communities except pure stands enjoy any appreciable degree of homogeneity. A purely homogeneous community would be a simple thing to study for theoretically one sample would assure a true picture of its character. Since variability of stand is to be expected in most communities, it is necessary to recognize the need for a number of samples. These are usually spaced at random so that these data lend themselves to statistical treatment.

Providing homogeneity of the community is great, the shape of the sample fails to be of consequence to the results of sampling. If the community is quite variable, better results may be obtained with some shape such as a rectangle where variation is sampled to better advantage. Size of a sample is also important as recognized by the fact that one would be more apt to find a larger number of species in a large sample than in a small one. Purpose of the investigator will largely determine size and shape of the sampling device.

The Quadrat

As young men, F. E. Clements and Roscoe Pound studied distribution of plants in their native state of Nebraska (1898). They devised a method for estimating percentages of different species comprising the communities of the prairie vegetation. Their method employed a square frame (quadrat) of one meter size which could be partitioned into 100 square decimeters. Vegetation thus could be observed in these decimeter areas and individual species estimated in proportion to each other. Estimations are based upon basal area of each species. In grassland studies this is grazing height. Basal area is a very stable characteristic of vegetation. It is estimated for each decimeter, too. All different kinds of plants in a decimeter are considered as 100 per cent. As each decimeter within the meter quadrat is examined, the composition is recorded for each on a data or record sheet (Fig. 18). Upon recording each of the 100 decimeter areas, sums of percentages for each species then yield their per cent composition. This procedure of obtaining sums of percentages is followed for each individual species. The grand sum of percentage by each species should total 100.

The method is at first somewhat tedious, but has the advantage of considerable detail and accuracy. Usually independent investigators are in close agreement on their figures. Of 100 square decimeters this automatically yields per cent composition for sums of individual species in the squares where each occurs. A few well-spaced quadrats make it possible to achieve by estimate the whole picture of community composition without counting each individual species over a wide area, which of course readily appears impossible.

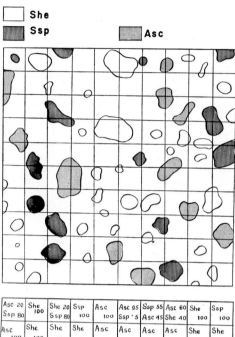

18. A representation of species distribution in a grassland quadrat. Estimations by species of the per cent each constitutes of 100 in each decimeter is made. A total is derived by summing figures for each species. The first two decimeter strips as recorded on a data sheet are shown below. For abbreviation, the capital letter of a genus and the first two letters (lower case) of each species are used. In this instance, the plants and their total composition in this quadrat are: prairie dropseed (*Sporobolus heterolepis*) 45.55 per cent; needle grass (*Stipa spartea*) 24.35 per cent; and little bluestem (*Andropogon scoparius*) 30.10 percent.

Line Interception

As mentioned, results of sampling are affected by size, number, and shape of quadrat. A rectangular sample area often is favored because it takes in a greater amount of heterogeneity of the stand. A method using this idea of longer sample is the line interception method (Canfield, 1941). It has indefinite length and can sample all layers by having as much vertical extension as needed. The only limited dimension is width.

The principle is to record all species which touch a line, made taut, a certain distance above the ground. This distance above ground will vary with vegetational type. In grassland communities it may be 1.5 inches. Coverage by diameter of stems or diameter of bunches along the line is measured and recorded to give figures used in calculating basal area, per cent composition, and per cent frequency of occurrence. One great advantage of the method is in its objectiveness. Experience has proved that interpretations are easier if the width is extended to 0.5 cm. of either side of the line or including all plants touching the line and within one full centimeter of one side of the line.

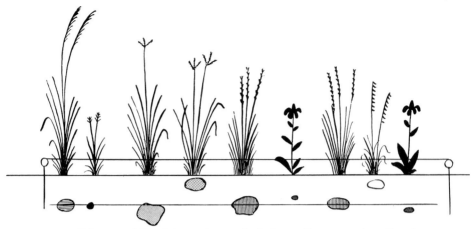

19. Diagram of line interception method of sampling vegetation. For obtaining total basal area the length of the line is divided by the length of intercepted areas of vegetation. In this instance total basal area is 33 per cent. Percentage composition is derived by dividing the length of intercepted areas of individual species by total length of all intercepted areas.

The line should be of heavy cord or wire and staked firmly into the ground (Fig. 19). Length of the line is decided somewhat by individual preference and vegetational type. Usually 10, 50, or 100 feet will be used. Because of its objectivity, it is less tiring and less tedious than subjective methods. For this reason, more samples may be taken in a shorter period of time. The method is applicable to all types of vegetation, including aquatics. In use with aquatic vegetation, some modification will have to be made for depth. The line may otherwise be stretched across open water and followed by boat or wader. Other modifications of the method may be used as long as the individual making the measurement remains the same and as long as the method is the same for different types being compared.

The Point "Quadrat" Method
(LEVY AND MADDEN, 1933)

The point "quadrat" method involves use of a linear transect meter which is mounted on legs or a frame. Ten metal pins (about 18 to 24 inches long) are inserted in holes of the one meter bar at intervals of one decimeter. The meter bar is mounted on the frame so that the pins approach the ground at an angle, or a vertical drop of the pins may be used.

The frame is placed in the vegetation to be sampled at regular intervals. Distance between samples will be determined more or less by the extent of the area of the type or community being sampled.

If the area is fairly extensive, then an interval of several paces (10 to 15 or 20) might be used. "Each contact of a point at the soil surface with the stem of a turf grass or forb or the crown of a bunch grass is recorded. Since the influence of shrub on the habitat is considered to be more important than is indicated by their basal areas, contacts of the pins of sampling apparatus with the branches and foliage of these species are recorded and not those made at the surface of the soil" (Coupland, 1950).

The results are expressed as percentage. Although no area is involved at a point, basal area can be calculated by recording the number of times a contact is made with the plant species in one hundred or one thousand projections. For example, if 142 contacts of vegetation are made in 1000 projections, then basal area would be 14.2 per cent. By recording contacts of point and individual species, basal area of each may be obtained.

Percentage composition may be calculated by considering the basal area of all species, whether large or small, as unity or 100 per cent. The percentage of this furnished by any one species would be considered as percentage composition. For example, in an area where the basal area of the species was 6 per cent and the basal area of the total vegetation furnished a figure of 12 per cent, the percentage composition of the one species would be 50 per cent. It is evident that from a lack of width dimension that a large number of samples might be necessary to insure an accurate picture of both basal area and percentage composition.

From the three previously described methods of studying vegetation, it will be easily recognized that each was developed for use in grassland vegetation primarily. The original meter quadrat was adapted to forest studies by making it a surveyor's chain length (66 feet) on a side (Lutz, 1930). This large quadrat was used in sampling the canopy species. The mid-layer of the forest was sampled by means of a quadrat 6.6' x 6.6' which was nested in one corner of the larger area. Understory species were sampled by means of a quadrat 3.3' x 3.3' and nested in one corner of the mid-layer quadrat. Each of the above areas used was a convenient fraction of an acre which facilitated shaping the data. Forest quadrats were later made ten square meters for canopy species and, because one size would not sample all layers properly, other adjustments in size were made for shrub and understory layers, too.

The methods of quadrat, line interception, and point quadrat or transect shows an evolution from a definite plot, square in shape, through rectangle to linear transect. The present product of this evolution of sampling device is the plotless method. Several such methods have been developed, each primarily for forest studies (Grosenbaugh, 1952; Cottam and Curtis, 1949, 1955, 1956).

Random Pairs Method
(COTTAM AND CURTIS, 1949, 1955, 1956)

One of the most recent methods of obtaining information concerning density, composition, frequency of occurrence of species, and other analytical characteristics of forest vegetation is the random pairs method. It has been used in the study of southern Illinois forest communities. This method is predicated on the assumption that trees of a forest vary randomly in distance from a theoretical condition of all trees being equidistant. Since spacing of individuals or distance is emphasized, the results of sampling will yield amount of area occupied by each plant. The "mean area" thus derived is a reciprocal of density (Phillips, 1959).

In a living forest where there is an actual random variation of distance between trees, the selection of trees in the random pairs method is made objective by use of a transect line and an exclusion (or inclusion) angle of 180 degrees. A compass line is usually established. Points along the line are set up by pacing off a predetermined interval. This interval should be great enough that a tree previously used will not be encountered again. In our woodlands, an interval of 20 to 25 paces works well. The two trees used in obtaining the distance (d) is tree A nearest the point P in figure 20, and tree B, the nearest one to it, but inside the 180 degree sector (as the observer faces tree A, this would be somewhere behind him). To be recorded are species A and B; their respective diameters at breast height providing they are over 4 inches; and the distance between the two trees. In using the 180 degree inclusion angle, it has been determined that the square of distance X 0.83 yields area occupied by a single tree. In this case 0.83 is a correction factor which must be used with the 180 degree angle.

Shrubs and young trees under 4 inches in diameter, but over 1 inch in diameter, are recorded for presence or number along a meter-wide transect between trees A and B. Herbaceous plants of the forest

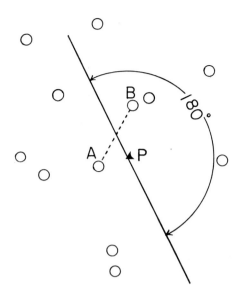

20. Hypothetical woodland situation in use of randomly selected pairs of trees. See text for explanation.

floor and young reproduction are recorded in quadrats of one meter at each point along the transect.

Using the random pairs method on trees over 4 inches diameter at breast height, the analytical characteristics of density, dominance, and frequency may be obtained from the following formulations:

1. Density (Trees/acre) $= \dfrac{43{,}560}{(0.83d)^2}$

   In this instance d refers to average of the distances encountered.

2. Relative Density (RD)

   RD of species A $= \dfrac{\text{number of individuals, species A}}{\text{total number individuals, all species}} \times 100$

3. Dominance (also may be called Per Cent Composition based on basal area)

   Dominance $= \dfrac{\text{total basal area of given species}}{\text{total basal area of all species}} \times 100$

   Our per cents of composition in all tables are thus based on basal area.
   Basal Area of single tree = ½ dia. of tree × 3.1417
   Total Basal Area (TBA) = Sum of basal areas of the trees under consideration.

   Relative Basal Area (RBA) $= \dfrac{\text{TBA of species A}}{\text{TBA of all trees examined}} \times 100$

   RBA is also used as Dominance.

4. Frequency (F) $= \dfrac{\text{No. of points in which a species occurs}}{\text{Total number of points examined}} \times 100$

   Frequency data in all tables presented have been derived in this manner.

5. Relative frequency (RF)

   RF of species A $= \dfrac{\text{Frequency of species A}}{\text{Sum of frequency value, all species}} \times 100$

PART III Plant Communities

The Lowland Series

South of the Shawnee Hills in southern Illinois in the Coastal Plain section (Mississippi Embayment) and in the Wabash Lowland there exist many acres of swamps, bayous, ditches, and wetlands. In the succession of aquatic vegetation water depth is usually considered a factor of prime importance. Stages of submerged, floating, reed-swamp, sedge-meadow, and woodland are usually in well-marked zonation and easily recognized (Weaver and Clements, 1938). In the secondary succession of woodland where more variables may exist, it is difficult to see the successional relationships. In addition to the factors of time and water depth are added aggradation and degradation, as well as human influences. The apparent stages of succession of woody vegetation in a wetland forest situation are as follows: Deep Swamp – Shallow Swamp – Transition Forest – Bottomland Forest. Penfound (1952) separates swamps into two categories: (1) "Deep swamps are fresh water, woody communities, with surface water throughout most or all of the growing season"; (2) "Shallow swamps are fresh water communities, the soil of which is inundated for only short periods during the growing season."

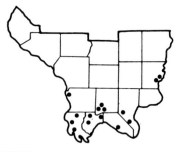

Deep Swamps

In constantly inundated areas or deep swamps the water level varies from about an inch or less at summer pool level to several feet at winter pool level. Quite often the summer pool level is as much as five feet deep. Trees which are most abundant in the deep swamps are bald cypress (*Taxodium distichum*), tupelo (*Nyssa aquatica*), water hickory (*Carya aquatica*), pumpkin ash (*Fraxinus profunda*), water locust (*Gleditsia aquatica*), and swamp red maple (*Acer drummondii*) (Fig. 21). Only rarely may be found the planer tree (*Planera aquatica*). The shrubby layer consists of swamp rose (*Rosa palustris*) and buttonbush (*Cephalanthus occidentalis*) or rarely water willow (*Decodon verticillatus*) (Figs. 22 and 23).

In deeper waters, the herbaceous plants which are characteristically found include: *Utricularia vulgaris*, pondweed (*Potamogeton diversifolius*), coontail (*Ceratophyllum demersum*), and naiad (*Naias flexilis*). A few free-floaters, many of them exceedingly rare

21. Trees of the deep swamps: (1) water locust (*Gleditsia aquatica*), (2) swamp rose (*Rosa palustris*), (3) pumpkin ash (*Fraxinus tomentosa*), (4) swamp cottonwood (*Populus heterophylla*), (5) tupelo gum (*Nyssa aquatica*).

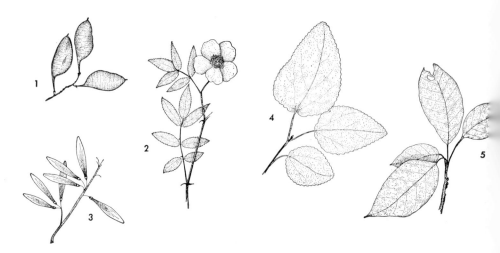

in Illinois, exist in more or less stagnant waters near the shore. Numbered among these are watermeals (*Wolffia columbiana* and *Wolffiella floridana*), duckweeds (*Lemna minor* and *L. perpusilla*), and *Spirodela polyrhiza*.

In the shallower portions of the deep swamps may be found several large but attractive herbs. These include water lotus (*Nelumbo lutea*), water lily (*Nuphar advena*), water-shield (*Brasenia schreberi*), sweet flag (*Acorus calamus*), arrow-arum (*Peltandra virginica*), bur-reed (*Sparganium eurycarpum* and *S. androcladum*), and sedges (*Carex lupuliformis, C. lurida, C. hystricina, C. lupulina,* and *C. stipata*).

Along shores are numerous rooted herbs. The swamp mallow (*Hibiscus lasiocarpus*) and species of beggar's-ticks (*Bidens* spp.) are common. Several sedges occur in clumps along the shores — *Carex frankii, C. squarrosa, C. typhina, C. vulpinoidea, Cyperus*

22. Shrubby species of the deep swamps: (1) duckweed (*Wolffiella floridana*), (2) buttonbush (*Cephalanthus occidentalis*), (3) water meal (*Wolffia columbiana*), (4) duckweed (*Lemna trisculca*), (5) duckweed (*Lemna minor*) (6) duckweed (*Spirodela polyrhiza*).

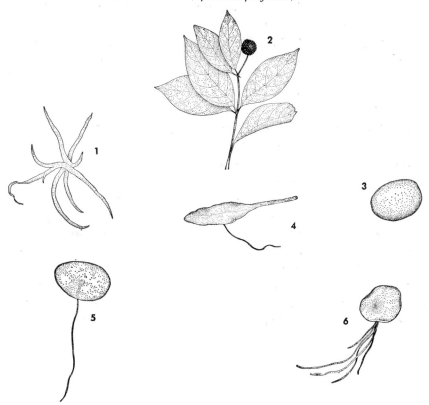

23. [left] Swamp rose (*Rosa palustris*) and buttonbush (*Cephalanthus occidentalis*), two shrubs of a deep swamp at LaRue, Union County. [right] Water willow (*Decodon verticillatus*) growing in the swamp at LaRue, Union County.

erythrorhizos, and *C. ferruginescens*. Occasional plants of water dock (*Rumex verticillatus*), aster (*Aster simplex* and *A. vimineus*), cardinal flower (*Lobelia cardinalis*), mild water-pepper (*Polygonum hydropiperoides*), and tear-thumb (*Polygonum sagittatum*) may be found. The species of shore grasses are fewer in number, but where grasses occur, they form dense stands. These shore grasses include species of cutgrass (*Leersia* spp.), munro-grass (*Panicum agrostoides*), and creeping bent-grass (*Agrostis palustris*).

While the *Taxodium-Nyssa aquatica/Rosa palustris* community was once fairly widespread in Alexander, Pulaski, Massac, Johnson, and Pope counties, drainage and subsequent clearing for cultivation have reduced this community to scattered local areas in these counties. In southern Illinois, typical *Taxodium-Nyssa aquatica/Rosa palustris* communities occur at the following stations: along Little Black Slough near Boss Island northwest of Belknap, Johnson County (sections 26 & 27, T 13 S, R 2 E); northwest of Karnak, Pulaski County (sect. 15, T 14 S, R 2 E); along Bay Creek ditch southeast of Renshaw, Pope County (sect. 5, T 14 S, R 5 E); along Cache River northwest of Pulaski, Pulaski County (sects. 4, 5, & 8, T 14 S, R 1 W); Horseshoe Lake and vicinity, south and west of Olive Branch, Alexander County. The future of the stations for this community is still perilous, although generally those which are left are very inaccessible.

24. [left] Bald cypress (*Taxodium distichum*) in a Pulaski County swamp. (Photo courtesy of Dr. Margaret Kaeiser.) [right] Cypress-knee sedge (*Carex decomposita*), a Coastal Plain species, grows upon cypress knees, buttresses, and logs of various woody species in swamps.

Bald Cypress (Taxodium distichum)

Bald Cypress is perhaps the most characteristic species of the deep swamps (Fig. 24). Together with tupelo gum it usually makes up nearly half the composition (Table 5). Cypress was, before cutting and draining of its native areas, one of the most conspicuous and outstanding trees of the swamps in the southern seven counties of Illinois. Today the bald cypress is still very abundant along the Coastal Plain of the United States.

Cypress reaches its greatest size in areas where water constantly inundates the land, but draw-downs are apparently needed for its reproduction. Seeds cannot germinate under water. Demaree (1932) concluded that seeds must germinate on emerged soil and reach a greater height than the next high water if they are to establish themselves. As a result this community often is composed of relatively old plants with little reproduction except along the periphery of the swamp. When conditions are favorable for growth, the seedlings develop rapidly.

The tallest Illinois cypress trees existing today are about 90 feet tall with a diameter of about 45 inches at breast height, although larger trees undoubtedly have grown in the past in southern Illinois. Brendel (1881) reports that the biggest tree he ever measured in Illinois "was a bald cypress, in Pulaski County. It had in four feet from the ground a diameter of 2.1 meters."

5. *Percentage of frequency and composition of tree species above 4" DBH (diameter at breast height) in a* Taxodium-Nyssa aquatica/Rosa palustris *community, Pulaski County*

SPECIES	Frequency	Composition
Taxodium distichum	19	27.5
Nyssa aquatica	20	22.0
Liduidambar styraciflua	9	11.0
Populus heterophylla	5	5.5
Fraxinus tomentosa	3	5.5
Salix nigra	5	5.5
Nyssa sylvatica	4	4.5
Carya aquatica	1	2.8
Ulmus rubra	9	2.8
Platanus occidentalis	4	2.8
Gleditsia triacanthos	5	2.8
Gleditsia aquatica	1	1.8
Betula nigra	2	1.2
Quercus palustris	2	1.0
Quercus macrocarpa	2	1.0
Celtis laevigata	1	1.0
Acer rubrum	1	1.0

Data from 54 random pairs. All areas of forest vegetation selected for study were at least 20 acres in size, showed no recent evidence of disturbance such as fire, grazing, or logging, and possessed a sufficient degree of homogeneity that subjective recognition of type was unmistakable.

Visible growth in cypress begins in late March or early April when needle-like leaves make their appearance. Simultaneously the male and female flowers are produced separately on the same tree. The male flowers are produced in purple clusters at the end of shoots of the preceding year. The small rounded female flowers, likewise borne on the preceding year's branchlets, may be found singly or in groups. By May, these delicate needles have attained their maximum length of ⅜ to ¾ of an inch. The fleshy, greenish-purple cones measure 1¼ inches in diameter and ripen during October. During September, the leaves begin to turn a dull brownish-red or even yellowish. In early October the needles fall, leaving the tree "bald" and exposing the beautifully excurrent shape formed by the branches.

Stalagmite-like structures protruding from the dark and still waters of the cypress swamp are "knees." They connect with the roots of the trees and are thought to function in the exchange of gases, particularly the supply of oxygen to the tree during times of prolonged flood stage. The waters are notably low in dissolved oxygen due to warm summer temperatures and the oxidation of organic matter. A very characteristic feature of the swamp where abundant knees are found is the presence of the southeastern cypress knee sedge (*Carex decomposita*) (Fig. 24). It grows epiphytically on the knees, bases of old trees, stumps, on fallen logs, or even upon fence posts which may protrude from the water.

Tupelo (Nyssa aquatica)

Tupelo gum is a typical Coastal Plain species inhabiting deep swamps. It is a common associate of bald cypress. As a characteristic of trees inhabiting flooded lands, the bases of the trunks of the tupelo gum are conspicuously swollen (Fig. 25). From this buttressed base arises a straight bole of some 55–80 feet in height. The bole bears a crown of small divergent branches. The trunk above the swollen base may have a diameter of 2.5 to 4.0 feet and is dark brown and rather deeply furrowed.

The leaves, formed in buds during earlier months, begin to unfold in April and become, at maturity, a shining dark green and average

25. Tupelo gum (*Nyssa aquatica*), a denizen of deep swamps, displays a trunk with pronounced buttressing in a Pulaski County swamp. (Photo by Dr. Lawrence Mish.)

6 inches long and about one-half as broad. The margins are usually entire, although occasional specimens may be found with a rather wavy or slightly toothed margin. The hairy petioles are about 1 to 2 inches long.

The usually unnoticed staminate and pistillate flowers are produced separately during April. The thick-skinned, dark purple fruits mature in September. Embedded in the exceedingly sour flesh is a flattened seed with winged ridges near the center. The fruits fall into the water at maturity and may be eaten by wildlife or they may float to shore where germination may occur. The tupelo is found from Virginia along the coast to Florida, westward to Texas, and north through the Mississippi Embayment into southern Illinois.

Taxodium-Fraxinus tomentosa/Itea virginica *Community*

The basic difference between this community and the *Taxodium-Nyssa aquatica/Rosa palustris* community is associated directly with the amount of standing water. Where water level stands at less than one foot, or the ground is at most only heavily saturated, the *Taxodium-Fraxinus tomentosa/Itea* community dominates. *Nyssa aquatica* gives way to *Fraxinus tomentosa* as the area becomes slightly higher in elevation. *Rosa palustris* is replaced by *Itea virginica* as the dominant species of the shrubby union.

This community is characterized by usually dense growth and a great diversity of species. Where this community occurs about five miles north of Dongola, twenty-one species of trees are present (Table 6). While *Taxodium distichum* and *Fraxinus tomentosa* are the most conspicuous elements, *Liquidambar styraciflua, Ulmus rubra,* and *Acer rubrum* occur almost as frequently. The Coastal Plain is represented in the woody flora of this community by *Taxodium distichum, Fraxinus tomentosa, Carya aquatica, Quercus michauxii* and, at some stations, *Gleditsia aquatica.*

A number of other shrubs participate in the *Itea virginica* union: *Planera aquatica,* the tallest of these, does little to determine the physiognomy because of its infrequent occurrence; it is most abundant along the Cache River between Perks and Rago; *Cephalanthus occidentalis* is scattered throughout the community; *Rosa palustris* is infrequent but occurs where water is deepest; *Styrax americana* is a rare shrub which occurs only near Rago.

Even though there is dense shading in this community, a well-

6. *Percentage of frequency and composition of tree species above 4" DBH in a* Taxodium-Fraxinus tomentosa/Itea virginica *community near Dongola, Union County*

SPECIES	Frequency	Composition
Taxodium distichum	18.6	11.6
Fraxinus tomentosa	20.9	11.6
Acer rubrum	20.9	11.6
Ulmus rubra	23.3	11.6
Liquidambar styraciflua	16.3	10.5
Acer negundo	13.9	7.0
Nyssa sylvatica	13.9	5.8
Acer drummondii	9.3	5.8
Carya laciniosa	9.3	3.5
Populus heterophylla	7.0	3.5
Quercus rubra	2.3	2.3
Quercus velutina	4.7	2.3
Celtis occidentalis	4.7	2.3
Quercus falcata var. pagodaefolia	4.7	2.3
Carya aquatica	2.3	1.5
Morus rubra	2.3	1.2
Quercus shumardii	2.3	1.2
Robinia pseudoacacia	2.3	1.2
Celtis laevigata	2.3	1.2
Acer saccharinum	2.3	1.0
Platanus occidentalis	2.3	1.0

Data from 43 random pairs.

developed herbaceous union with a preponderance of sedges and grasses occurs. Robust species of *Carex* and tall grass make this a conspicuous layer. *Carex grayi* and *C. louisianica* are abundant, while *C. muskingumensis* is found frequently. *Cinna arundinacea* is a leading species of grass in clearings and wood's margins, though *Leersia oryzoides* and *L. lenticularis* also are found in considerable abundance in openings and marginal situations. Areas where this community is typical in southern Illinois are: Cache River bottoms near Rago, Johnson County; east of Dongola, Union County; between Pulaski and Cypress, Pulaski County.

Pumpkin Ash (Fraxinus tomentosa)

Pumpkin ash, another Coastal Plain species, has limited occurrence with cypress and tupelo in the deep swamp. It develops best where the waters are not as deep as for bald cypress and tupelo gum, yet water depth usually varies from about a foot deep to ground which is only heavily saturated.

The trunk is straight and the buttressing again somewhat pronounced. The leaves are compound, having seven leaflets on a woody petiole. They sometimes attain a length of 20 inches and are yellow-green color. The flowers, borne in elongated crowded racemes, are produced in late April and early May. These flowers are without petals and are dioecious. The fruits are linear-oblong, 2 to 4 inches long; the wing is one-quarter to one-half inch broad. The fruits become brownish and ripen in late summer or fall.

Exceptionally large trees reach a height of 100 feet in the swamps of southern Illinois. Miller and Tehon (1929) report this species to attain a height of 69 to 98 feet and a diameter at breast-height of 16 to 26 inches in Pulaski County. The bark of pumpkin ash is dark gray, shallowly fissured into broad, scaly ridges. The wood lacks the strength and toughness which makes white ash a valuable timber tree. It finds its principal use in crating, boxes, paper pulp, or fuel. The distribution of this tree in the United States is Indiana, Illinois, south to the Gulf of Mexico and to Florida.

Fraxinus lanceolata-Populus heterophylla/Cephalanthus Community

This unique, distinctive swamp community is limited in Illinois to the Mississippi border of Union and Alexander counties and to the Lower Wabash border on the opposite side of the state.

In these communities, water stands in some portions throughout the year. North of Wolf Lake (including Wolf Lake Swamp and La Rue Swamp), the water level is stabilized by springs whose waters maintain a temperature of about 57° F. throughout the year (Gunning and Lewis, 1955).

The absence of *Taxodium distichum* (except for a few trees in the deepest parts of the swamps) and *Nyssa aquatica* is conspicuous along with the reduced numbers of *Fraxinus tomentosa*. Aside from the two main dominants, other important species of the tree union are *Gleditsia aquatica* and *Celtis laevigata*. Many dead trees stand in the swamp as testimony to their intolerance of prolonged standing water. The occurrence of these dead trees permits the water to be well-lighted. This commonly allows the shrubby constituents of the *Cephalanthus* layer to increase in size and numbers until a group of tree seedlings assumes dominance. *Cephalanthus occidentalis* comprises nearly 90 per cent of the shrub layer in waters three to six feet deep, but where water is shallower, *Itea virginica, Decodon verticillatus,* and *Rosa palustris* are common.

Green Ash (Fraxinus lanceolata)

Green ash, a tree confined mostly to bottomland situations, often reaches heights of 50 to 60 feet. A record individual 5 feet in diameter and 138 feet in height is reported from the Wabash Valley (Wright, 1959). Leaves are compound with 5–9 leaflets. Leaflets are bright green on the under surface and this is in contrast to the white under surface of leaflets of white ash (*Fraxinus americana*). Field identification of these two species is often difficult because leaves may be beyond reach or fruits are not available. Because of this frequent difficulty and the lowland habitat, we have not distinguished them in the statistical data. When fruits are available, the distinction is easy. Green ash has a fruit in which the samara, or wing, is decurrent along half the seed length. In white ash, the samara is not decurrent along the seed but appears to be at the end of the seed.

Green ash is dioecious which is to say that some trees bear staminate flowers and others bear pistillate flowers. Flowers develop in April or May, and this occurs before leaves are unfolded. The period of pollination is brief, usually only 3 to 4 days, and pollen falls over an area of but 200 to 300 feet from the source. Pistillate flowers on other trees enlarge and open a few days later than the male flowers. Stigmas are immediately receptive to pollen, and samaras develop from the fertilized flowers within a month. Growth of embryos is not complete until September or October. Seeds fall through winter and spring and are dispersed by wind over a distance up to a hundred feet (Wright, 1959).

Green ash is a pioneer species in succession. It may follow willow and cottonwood. It is adapted to flooding, and young trees often put forth adventitious roots at the mark of high water on the stem (Fig.

26. [left] Adventitious roots on a young green ash mark a former high-water level. [right] A swamp community of box elder and silver maple. The undergrowth is mainly seedlings of silver maple.

26). Intense shading in late stage communities is a condition which is deleterious to green ash. The ash does not grow as fast as maples and elms and therefore it is unable to maintain its canopy position. During 1958, a flood year in the Wabash valley, green ash showed only 1 to 1½ inches of twig growth while nearby specimens of hackberry showed 12 inches of elongation and shagbark hickory 5 inches of twig elongation. High water marks on these trees during this year of growth was about 5 feet and the period of duration two to three weeks, recurrent several times from spring to late summer. Most commonly associated species are also pioneers such as cottonwood, willow, box elder, silver and red maples, pin oak, and sweetgum (Fig. 26).

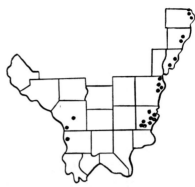

Bottomland Forests

In areas of partial inundation during the year, a quite different plant community develops. The dominants of this situation are black willow (*Salix nigra*), cottonwood (*Populus deltoides*), swamp cottonwood (*Populus heterophylla*), red maple (*Acer rubrum*), honey locust (*Gleditsia triacanthos*), and, in lesser number, silver maple (*Acer saccharinum*), box elder (*Acer negundo*), slippery elm (*Ulmus rubra*), and American elm (*Ulmus americana*). Only a few shrubs constitute the understory—gray dogwood (*Cornus obliqua*), swamp privet (*Forestiera acuminata*), hawthorn (*Crataegus* spp.), and southward, the Virginia willow (*Itea virginica*).

The understory is often sparsely populated, but a large number of species occurs. Characteristic plants of the periodically wet banks are *Polygonum pennsylvanicum, Polygonum persicaria,* rushes (*Juncus* spp.), and sedges(*Carex frankii, C. squarrosa, C. typhina,* and *C. crus-corvi*). The characteristic plants of periodically wet depressions include swamp milkweed (*Asclepias perennis*), cardinal flower, and ditch stonecrop (*Penthorum sedoides*). In depressions which are wet for the longest periods may be found lizard's-tail (*Saururus cernuus*). Other plants of the understory are pink st. john's-wort (*Triadenum walteri*), hydrolea (*Hydrolea affinis*), and monkey-flower (*Mimulus ringens*). In areas of less prolonged flooding there develops occasionally a community of cane (*Arundinaria gigantea*). In fact, cane usually may be found just beyond the high water level.

Black Willow (Salix nigra)

The generic name *Salix* takes its meaning from the inseparable association of willow and water's edge (Eifert, 1959). At river's edge nearly everywhere, dating back to the Cretaceous period and perhaps older than "Old Man River," is the willow. The success of this plant in meeting the perils of a fluctuating river bank environment is found in its structure and physiology. The long slender twigs bend gracefully and with full sweep during windy weather without breaking. The slender, elongated, glossy leaves with their reduced friction likewise escape tearing. The willow transpires a tremendous quantity of water and thrives in the poorly aerated soils of the riverside. The period of seed viability is short. Usually after two or three weeks, if suitable conditions have not been obtained, there will be no germination. For those seeds falling into water or into a shallow depression within the short span of viability, there will be germination within a few days, either on or in the water. The seedlings are tolerant to submergence for about a month (Hosner, 1957). Mature willow trees have lived for eight years, longer than other bottomland species, under continuous standing water (Yeager, 1949). Because of its early spring growth and flowering during cooler parts of the year, a northern ancestry is suspected.

Catkins, the small clusters of unisexual flowers, appear in early spring before leaves appear. The black willow is dioecious. Pollen is produced in abundance. Its transfer is accomplished by both bees and wind. A copious amount is wasted over the water surface and upon mudbanks where it falls in a golden shower. Soon, from pistillate catkins, are produced seeds which erupt and ride the wind to moist banks and shallows, or they fall upon water.

Black willow, our commonest species, often forms nearly pure stands or mingles with its family kin, the cottonwood (Table 7). Few shrubs and only limited tree species are found in the type. Grape and trumpet creeper are the commonest vines. Rice cutgrass occupies low, open places in the understory. Low places where the understory is more shaded may be occupied by *Carex typhina*. This species has been observed to survive four weeks of submergence during natural flooding and has lived through sixty days of artificial submergence. Upon removal from flooded condition, its growth and flowering were approximately normal. Scatterings of aster, violets, *Bidens, Lindernia,* and smartweed (*Polygonum pennsylvanicum*) usually may be encountered.

THE LOWLAND SERIES

7. *Percentage of frequency and composition of tree species over 4" DBH in a* Populus deltoides-Salix nigra/Leersia *community in the Lower Wabash Valley, Auroras Bend*

SPECIES	Frequency	Composition
Populus deltoides	35	57.1
Salix nigra	24	25.0
Acer saccharinum	12	12.3
Gleditsia triacanthos	7	1.4
Ulmus americana	3	1.6
Fraxinus americana	7	2.7

Data from 50 random pairs.

Cottonwood (Populus deltoides)

Chance dissemination is largely responsible for development of a cottonwood or black willow community. These two species of the family *Salicaceae* share the similarities of habitat, adventitious root production, seed viability period, and germination conditions. It is understandable then that they are found so often co-mingling along the river's edge.

Eastern cottonwood may reach a height of nearly 100 feet. It has soft wood and makes rapid growth. Actually, it often grows faster than such tall lowland herbaceous species as sunflower and ragweed. Its leaves are glossy, triangle-shaped, and attached to stems by flattened petioles which cause a tremoring effect in light breezes.

Small greenish flowers are produced in catkins in early spring (mid-March). Staminate and pistillate flowers are borne in separate catkins, each type on a different tree. Soon after flowering, the cottony or comose seeds are produced. During dry breezy early summer days, these fruits ride the wind currents. Like willow, they may be wafted to a moist bank or shallow, may float, then sink and germinate. High flood marks on banks are shown by lines of seedlings when water has subsided. The viability period, too, is only a couple of weeks and prolonged soaking seems deleterious to its germination (Hosner, 1957).

Commonest understory plants of nearly pure cottonwood stands of pole size include goldenrods, trumpet creeper, Illinois mimosa (*Desmanthus illinoensis*), violets, marsh elder (*Iva ciliata*), poison ivy, and smooth buttonweed (*Spermacoce glabra*). These last two species are usually more abundant than others. There may be seedlings of elm, maple, hackberry, and sweetgum among the understory herbs.

Hawthorn and deciduous holly are shrubs appearing occasionally. Where willow and cottonwood are co-dominant, the principal herbs may be *Leersia* spp. and *Aster simplex* (Table 8).

8. Percentage of composition, based on foliage cover, and frequency of a Populus deltoides-Salix nigra/Leersia understory community, Auroras Bend, Wabash Valley

SPECIES	Frequency	Composition
Acer saccharinum seedlings	32	59.6
Leersia spp.	16	22.4
Aster simplex	6	7.8
Polygonum pennsylvanicum	2	3.0
Vitis spp.	1	2.2
Amaranthus spp.	1	1.7
Populus deltoides seedlings	1	1.3
Agastache nepetoides	2	0.6
Lindernia dubia	1	0.4
Pilea pumila	1	0.4
Gratiola neglecta	1	0.2
Viola spp.	1	0.2

Data from 50 one-meter quadrats.

The Lower Wabash

Forest vegetation of river valleys is at present highly residual. These areas have been cultivated extensively because of their high yields of cash crops. Cultivation often has extended directly to the banks of levees and the only places of little disturbance are in the sloughs. The best early day account of river valley vegetation, in primeval condition, is that of Ridgway (1872; 1872b) who wrote of the conditions in the Lower Wabash Valley.

This area long has been recognized as a place of unusual botanical interest. The Coastal Plain element is an easily recognized part of the valley flora. Peattie (1922) listed 207 species, varieties, and forms of Coastal Plain origin in Indiana. Forty-two per cent of this number is found in the Ohio and Wabash River counties. Deam (1940) points out the characteristic ligneous species of Indiana which are located in the valley: *Acer drummondii, Carya illinoensis, Celtis laevigata, Forestiera acuminata, Gleditsia aquatica, Gleditsia texana, Taxodium distichum,* and *Quercus lyrata.* Jones, et al. (1955) list the characteristic ligneous plants of the Illinois side of the valley and include, in addition: *Acer saccharum, Tilia americana, Liriodendron tulipifera,*

THE LOWLAND SERIES

Quercus falcata var. *pagodaefolia, Quercus shumardii,* and *Liquidambar styraciflua.* These are not exclusive to the Wabash Valley but show a northward extension mostly in the river valley. They may be described best as Mississippi Valley plants finding their northeastern limits in this area. To those species listed by Jones, *et al.,* may be added *Fagus grandifolia, Prunus munsoniana,* and *Catalpa speciosa.*

Herbaceous Mississippi Valley plants listed by Deam (1940) are *Aristolochia tomentosa, Echinodorus radicans, Hottonia inflata, Ludwigia glandulosa, Spigelia marilandica, Trachelospermum difforme,* and *Vitis palmata.*

In the Illinois part of the valley, another group of herbs finds its northeastern limit. Among this may be listed: *Arundinaria gigantea, Carex typhina, Cyperus densicespitosus, Commelina virginica, Phoradendron flavescens, Hydrophyllum macrophyllum, Scutellaria elliptica, Orobanche uniflora, Bignonia capreolata, Valeriana pauciflora, Helianthus microcephalus, Mikania scandens,* and *Silene regia.* (For typical examples see Fig. 27.)

Ridgway (1872) writes of the rich ligneous flora of the Lower Wabash Valley: "In this section of the country many species of the bo-

27. Some plants of the bottomland forests: (1) mistletoe (*Phoradendron flavescens*), (2) sweet gum (*Liquidambar styraciflua*), (3) swamp privet (*Forestiera acuminata*), cherrybark oak (*Quercus falcata* var. *pagodaefolia*), (5) featherfoil (*Hottonia inflata*).

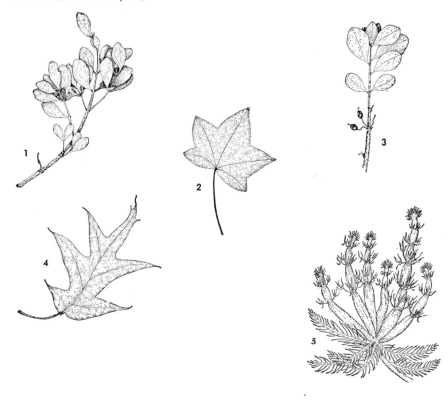

tanical districts named, in receding from their several centers of abundance, overlap each other, or reach their latitudinal or longitudinal limits of natural distribution. . . . The mixed woods of the Lower Wabash Valley consist of ninety species of trees, including all those of which reach a maximum height of over twenty feet: these are distributed among about twenty-five families and fifty genera. In the heavy forests of the rich bottom lands more than sixty species grow together, though in various localities different species are the predominating ones . . . in the rich alluvial bottoms, the deep soil . . . nourishes black walnuts, tulip trees, sycamores, white ashes and sweetgums of astonishing dimensions. Trees which usually attain largest size are the following species, named in the order of their maximum size: sycamore (*Platanus occidentalis*), tulip poplar (*Liriodendron tulipifera*), pecan (*Carya illinoensis*), overcup oak (*Quercus lyrata*), etc.

"The river flows for the greater part between dense walls of forest, which stand up to the very banks and generally screened in front with a dense fringe of willows, with a belt of cottonwood and sycamores behind it . . . the fronting belt of willows which are often overrun by luxuriant masses of wild grape or other vines, often falling down to the very waters edge." Viewed from the tops of bluffs, Ridgway says: "It appears a compact level sea of green, apparently endless, but bounded by the line of wooded bluffs 3 to 7 miles back from the river.

"Going into these primitive woods, we find symmetrical, solid trunks of six feet and upwards in diameter, and fifty feet or more long to be not uncommon, in a half dozen or more species; while now and then we happen on one of these old sycamores, for which the rich alluvial bottoms of the western rivers are so famous, with a trunk thirty or even 40, possibly 50 or 60 feet in circumference. . . . The tall shaft-like trunks of pecans, sweetgums or ashes, occasionally break on the sight through the dense undergrowth or stand clear and upright in unobstructed view in the rich wet woods, and rise straight as an arrow for eighty or ninety, or perhaps over a hundred feet before the first branches are thrown out."

On the structure of the forests, Ridgway (1872b) says: "and going into the woods, we found them to be almost completely primitive in their condition, and so dark and silent that one could easily imagine himself in a wholly uninhabited region. . . . The largest trees were of course the gigantic sycamores with trunks twenty-five—thirty feet in circumference. . . . Nowhere else had we seen the sweetgum growing in greater abundance and to such magnificent proportions. In the damper parts of this forest it formed the prevailing growth and seemed to vie with the majestic pecan in its towering height . . . branches of American elm and honey locust were matted with mistletoe, which plant evinces in this region a striking partiality to these trees."

The understory was described by Ridgway as being composed chiefly of pawpaw, red mulberry, sassafras, redbud, blue beech, flowering dogwood, and black haw. The undergrowth was described often as being too dense to penetrate and composed of *Zanthoxylum, Ptelea, Staphylea, Euonymous atropurpureus, Crataegus spathulata* [*C. crusgalli*], and *Lindera*. In hollows parallel to the river the small cane formed dense brakes, and grew 10 or 12 feet high, the canes matted with thorny 'green-brier' and mixed with tall stinging nettles (*Urtica* and *Laportea*)."

Particular mention (Ridgway, 1872; 1872b) is made of vines and their variety and size: *Vitis cordifolia* [*V. vulpina*] having circumferences of 20 to 40 inches and poison ivy growing up the trunks of trees in which one specimen measured 41 inches in circumference several feet from its roots were noted. The following quote from Ridgway (1872) enumerates the more common species of vines: "Old decaying trunks, on every hand, were encased in thick matted covering of Virginia creeper. Smaller vines, as *Cocculus carolina, Menispermum canadense, Celastrus scandens*, green-briers (*Smilax rotundifolia, S. glauca, S. tamnoides* [*S. bona-nox*], *S. walteri* [?], *S. lanceolata* [?], and perhaps one or two other species), pipe vine (*Aristolochia*), and many others screened the shrubbery or festooned the underwood, while a great variety of herbaceous vines, far too numerous to name in full, trailed over the undergrowth or ran up the shrubbery. Chief among these were the virgin's bowers (*Clematis pitcheri, C. viorna*, and *C. virginiana*), the yellow passion flower (*Passiflora lutea*), wild cypress vine (*Quamoclit coccinea*), wild blue morning glory (*Ipomoea* spp.), rutland beauties (*Convolvulus sepium* and *C. spithameus*), balsam apple (*Echinocystis lobata*), wild hop (*Humulus lupulus*), wild yam (*Dioscorea villosa*), and carrion flower (*S. lasioneura*). Besides there were several species of dodder (*Cuscuta*) which spread a carpet of orange-colored yarn, as it were." Five species of vines were observed by Ridgway (1872) climbing upon

9. *Vines encountered on trees, one-mile transect in bottomland woods near Palestine, Illinois*

SPECIES OF VINE	Number of Trees	Percentage of Total Vines
Rhus radicans	57	52.7
Vitis spp.	37	31.4
Menispermum canadense	8	7.4
Smilax spp.	6	5.5
Campsis radicans	3	2.7

one tree. These were *Rhus radicans, Campsis radicans, Smilax rotundifolia, Celastrus scandens,* and *Menispermum canadense.* Even today, vines are an outstanding characteristic of the residual forests of the Wabash lowland. During field studies in the upper part of the valley, poison ivy and grape were found to be the commonest vines growing upon trees (Table 9).

Along floodplains where water remains for a longer time after each rainfall, certain hydrophytic species such as swamp buttercup (*Ranunculus septentrionalis*), cordate plantain (*Plantago cordata*), species of *Polygonum, Carex* spp., numerous grasses, and others may be found. Where currents are rapid, little vegetation is able to anchor itself so that the flora associated with a fast growing stream is sparse, save for some of the lower plants, such as algae and mosses. As the energy from the stream is checked, conditions for plant life become more favorable. Typical species to be found on gravel beds and sandy shores of southern Illinois streams are *Commelina virginica, Lindernia dubia, Linum striatum, Hypericum mutilum, Lobelia cardinalis, L. siphilitica,* and a few grasses and sedges. Occasional shrubs such as *Salix sericea* and *Alnus serrulata* find the banks of streams ideal habitats.

The vegetation of newly formed sandbars usually contains weedy species (*Ambrosia trifida, A. elatior,* and *Xanthium chinense*), along with *Juncus bufonius, J. tenuis, Eleocharis acicularis,* and other small rushes and sedges. Soon black willow and sandbar willow (*Salix nigra* and *S. interior,* respectively) and cottonwood (*Populus deltoides*) appear on the bars. Herbaceous species become established near the water and these may be found on nearly every sandbar. These include *Eragrostis hypnoides, Leptochloa filiformis, Hemicarpha micrantha, Rumex fueginus, Potentilla paradoxa,* and species of *Bidens.*

Floodplains

Width of present day floodplains and immediate terraces vary from very narrow strips to wider expanses of several hundred yards or in places to one or two miles wide. These wider expanses of floodplains may more properly be termed swamp phases of bottomland forests such as may be presented in the remaining cutoffs and sloughs.

Communities representing stages of development in all bottomland hardwood forests are related developmentally through decreased hydrophytism. An increased shading in later vegetational development gradually replaces the degree of soil saturation as a dominant influence. In the more strictly aquatic series, dominance of a factor or set of factors is fairly obvious. These correlations of developmental sequence with site factors are obscured in the transition forests. Here are more species and a topography so gently expressed as to present a factor interaction difficult to correlate on a topographic basis alone.

These site quality differences, however, may be judged on other bases than topography. Among those which may be mentioned are soil depth and moisture, parent material, flora, and plant growth. Plant growth integrates and expresses all these in a single measurable condition. Hence the more we know of all plants in their environments, particularly in terms of their optimal and limiting conditions, the better judgment of site conditions we can make.

Understory and Site Condition

Herbs are an integral part of the structure of woodlands. They are principal constituents of the forest understory and share this layer with seedlings of the canopy and mid-layer. They are the familiar beauty and a distinctive part of woodland, often strikingly so as in the case of unmixed thickly aggregated plants of a particular seasonal aspect. They contribute much to the physiognomy or "look" of a stand. This has been demonstrated in our description of plant communities.

By observing occurrence of certain herbs with particular dominants and by knowing the relation of these species to drainage or shade condition, we may use them as site indicators. Light controls the height or structure of the undergrowth, but it is moisture which largely controls composition.

There are many species which enjoy a wide ecological amplitude. These are found in a variety of habitats. They are ubiquitous species and are unsuited as indicators. Other plants are more specific in their

10. *Vegetation-moisture table*

Cottonwood, Willow, Birch

HEAVY-WET (LITTORAL-AMPHIBIOUS)

Lizard Tail (Saururus cernuus), Arrow-Arum (Peltandra virginica), Water Plantain (Alisma plantago-aquatica), Sedge (Carex muskingumensis), Marsh Fern (Dryopteris thelypteris), Bur-Reed (Sparganium eurycarpum), Water Privet (Forestiera acuminata), Rice Cut Grass (Leersia oryzoides), Spike Rush (Eleocharis obtusa), Buttonbush (Cephalanthus occidentalis)

WET (OVERFLOW BOTTOM)

Cane (Arundinaria gigantea), Sedge (Carex typhina), Fowl Manna Grass (Glyceria striata), Deciduous Holly (Ilex decidua), Hawthorn (Crataegus spp.), Swamp Milkweed (Asclepias perennis), Spider Lily (Hymenocallis occidentalis), Sedge (Carex grayi), St. John's Wort (Hypericum punctatum), Ditch Stonecrop (Penthorum sedoides)

Red Maple, Sweetgum, Pin Oak

WET-MOIST (OVERFLOW TERRACE)

Grape Fern (Botrychium obliquum), Bittersweet (Celastrus scandens), Coralberry (Symphoricarpos orbiculatus), Trumpet Creeper (Campsis radicans), Sedge (Carex squarrosa), Wood Nettle (Laportea canadensis), Virginia Knotweed (Polygonum virginianum), Catnip (Agastache nepetoides), Hedge Hyssop (Lindernia dubia), Sensitive Fern (Onoclea sensibilis), Nettle (Urtica gracilis), Mist Flower (Eupatorium coelestinum), Raccoon Grape (Ampelopsis cordata)

Sycamore, Elm

VERY MOIST (AGGRADED TERRACE)

Wood Reed (Cinna arundinacea), Loosestrife (Lysimachia lanceolata) Jewel Weed (Impatiens biflora), Hemlock (Cicuta maculata), Clearweed (Pilea pumila), Smartweed (Polygonum pennsylvanicum), Wood Hyssop (Stachys tenuifolia), Wild Senna (Cassia marilandica), Elderberry (Sambucus canadensis), Elephant's Foot (Elephantopus carolinianus), Wahoo (Euonymus atropurpureus)

Walnut, Pecan

MOIST

Sea Oats (Uniola latifolia), Ruellia (Ruellia strepens), Wild Rye (Elymus virginicus), Sedge (Carex caroliniana), Indian Pink (Spigelia marilandica), Land Starwort (Callitriche terrestris), Pitcher's Clematis (Clematis pitcheri), Climbing Bean (Apios americana), Crownbeard (Verbesina alternifolia), Germander (Teucrium canadense), Monkey Flower (Mimulus alatus), Wild Yam (Dioscorea villosa), Prickly Sida (Sida spinosa)

CHIEF UBIQUITOUS SPECIES: Poison Ivy (Rhus radicans), Aster (Aster ontarionis), Violet (Viola spp.).
RARE PLANTS: Carolina Snailseed (Cocculus carolinus), Heliotrope (Heliotropium indicum) Meadow Parsnip (Zizia aurea).

THE LOWLAND SERIES

requirements and are found consistently in their more or less particular situations. These are very usable indicators. Still other plants may be found in more than one situation and are still not ubiquitous. By observing these herbs in the places of their optimum development, they also may serve as indicators.

It is known generally that extreme habitats present fewer species than mesic ones. Thus aquatic or wet-land plants are more confined to their habitats than those of better drained areas. To put it another way, these plants are more faithfully present in samples within the same habitat type than some of these species of more mesic situations.

In two years of field observation in the Lower Wabash valley, a vegetation table (sensu Rowe, 1956) was prepared by placing species exclusive of ubiquitous and rare ones into five moisture classes (heavy-wet, wet, wet-moist, very moist, and moist). These divisions reflect the relative drainage situation. The relation of tree species to the moisture gradient is shown at the top of Table 10. It will be re-

11. Acer saccharinum-Betula nigra/Saururus *community*

ENVIRONMENT	Species
Heavy-Wet	Saururus cernuus, Peltandra virginica, Leersia spp., Carex muskingumensis, Sparganium spp.
Wet	Dryopteris thelypteris, Polygonum virginianum
Wet-Moist	Arisaema dracontium

$(5 \times 16) + (2 \times 8) + (1 \times 4) = 100/(5 + 2 + 1) = 12.5 \times 10 = 125$

Drainage Index 125.

12. Quercus alba-Carya ovata/Uniola latifolia *community*

ENVIRONMENT	Species
Wet	Hymenocallis occidentalis
Wet-Moist	Symphoricarpos orbiculatus, Carex caroliniana, Carex spp., Campsis radicans
Moist	Ruellia strepens, Uniola latifolia, Lysimachia ciliata, Arisaema dracontium, Spigelia marilandica, Callitriche terrestris, Clematis pitcheri, Hedeoma pulegioides, Euphorbia corollata, Polygonum scandens

$(1 \times 8) + (4 \times 4) + (9 \times 1) = 33/(1 + 4 + 9) = 2.4 \times 10 = 24$

Drainage Index 24

membered that some plants grow in more than one community type, but an attempt has been made to place each into a moisture class which apparently is an optimum one for it. The presence of certain herbs in a particular part of the table for a stand shows the importance of these plants in the physiognomy of that particular community (Tables 11 and 12).

A drainage index was calculated in the same manner that Rowe (1956) calculated his vegetational moisture index. The procedure will be briefly described. The number of plants listed for the heavy-wet class was multiplied by 16, the wet area plants by 8, wet-moist number was multiplied by 4, very moist by 2, and the number of moist area plants was multiplied by 1. This geometrical increase in weighting gave a scale of values with a nice spread and also recognized the significance of the wet habitat and close association of a certain small number of plants. Fifty widely spaced quadrats rarely produced a list of herbaceous plants greater than twenty in number from lowland habitats. After removal of ubiquitous species and rare plants, there often were just a few for calculation of a drainage index.

In nine bottomland forests of the Lower Wabash valley, the Drainage Index range was 125, 106, 88, 66, 54, 47, 40, and 24. The communities by this ordination are shown with their composition figures in Table 13. Since drainage is the outstanding factor of the environment of these floodplains, it is presumed that this represents the successional pattern among the dominants. In a study of 18 bottomland areas of the Lower Wabash valley, there were 134 herbaceous species encountered in 831 meter quadrats examined on the forest floor. Only those of greatest ecological significance are presented (Table 14.)

The large number of species (19) having total combinations less than 3 times with other species and the large number of combinations among the more abundant species emphasize a poorly expressed community. Pecan, having a combination with itself 17 times, probably has been favored because of its value as a nut tree. The large number of combinations seems not entirely a natural thing. Pin oak combined with itself 9 times which was highest for combinations of a tree with itself in a situation regarded as entirely natural. Pin oak-dominated areas easily are recognized throughout the valley, and it constitutes one of the most easily recognized communities. Pin oak also often combines with sweetgum.

An analysis of the understory shrubs and tree reproduction of a

13. Percentage composition based on basal area of bottomland hardwood forests of lower Wabash Valley

SPECIES	1	2	3	4	5	6	7	8	9
Quercus alba	31.9	10.8	....	....	....	15.8	10.1	....	4.1
Carya ovata	17.6	....	....	1.5	2.6	1.8	17.5	....	....
Quercus lyrata	16.7	....	....	10.5	0.2	....	1.4	....	10.5
Quercus velutina	11.8	19.0	....	1.4	....	3.6	13.4	....	1.2
Quercus palustris	10.9	12.3	3.6	1.7	18.1	6.4	1.1	0.4	....
Fraxinus pennsylvanica var. lanceolata	0.9	....	1.7	0.2	7.0	5.8	1.4	5.6	2.8
Quercus bicolor	0.7	....	0.3	....	6.0	11.3	1.8	0.8	1.9
Gleditsia triacanthos	0.2	0.4	1.2	1.4	0.9	6.4	....	....	0.2
Ulmus americana	0.2	....	32.8	....	....	18.8	9.0	6.5	....
Carya illinoensis	....	17.2	....	41.8	15.7	1.8	1.1	....	3.2
Liquidambar styraciflua	....	6.5	6.5	....	9.5	....	8.8	2.6	16.0
Acer saccharinum	....	4.6	2.6	0.3	1.6	2.4	....	27.4	....
Platanus occidentalis	....	3.7	7.3	1.6	....	0.4	....	16.6	....
Juglans nigra	....	3.3	12.3	....	4.0	3.1	....	0.5	....
Ulmus rubra	....	3.2	4.2	10.6	8.0	....	3.9	1.3	....
Nyssa sylvatica	....	1.8	....	2.2	....	....	6.4	0.4	2.9
Quercus macrocarpa	....	1.6	7.4	....	1.6	1.5	5.9	5.5	0.2
Celtis occidentalis	....	1.1	5.0	....	3.2	0.9	1.6	....	....

Each community based upon a minimum of 100 trees, 50 random pairs. Areas represented: (1) North Fork of the Saline River, Saline County; (2) New Haven, Ill.; (3) near Calvin, Ill., Beaver Crossing; (4) Little Wabash near Emma, Ill.; (5) south of Mt. Carmel, Ill.; (6) near Calvin; (7) east of Inman, Ill.; (8) cattail slough south of New Haven; (9) north of New Haven. In addition to the above species, the following constituted the percentages indicated in from one to four areas:

Four areas: Carya cordiformis (0.8–8.8), Carya glabra (0.2–3.8), Populus deltoides (1.2–10.7), Acer negundo (0.2–3.0), Carya laciniosa (2.0–13.2), Crataegus spp. (0.2–0.4).

Three areas: Catalpa speciosa (1.0–4.8), Quercus rubra (0.7–8.5), Acer saccharum (2.0–13.3), Betula nigra (1.0–18.7), Acer rubrum (3.9–13.7), Salix nigra (1.5–5.5), Quercus imbricaria (0.3–3.9), Celtis laevigata (0.4–3.7), Cercis canadensis (0.1–0.4).

Two areas: Diospyros virginiana (0.1–0.4), Populus heterophylla (0.2–1.0), Tilia americana (0.2–0.6).

One area: Ostrya virginiana (0.9), Fraxinus americana (0.7), Juglans cinerea (0.3), Catalpa bignonioides (0.5), Gymnocladus dioica (1.9), Sassafras albidum (0.2), Quercus muhlenbergii (0.1), Carya tomentosa (0.5), Morus rubra (0.8).

forest will give clues as to future generations of dominant trees and understory. Correlation also may be made between small numbers of individuals in different size classes with observed habitat influences or reactions. Arm-wide transects (381) between randomly selected trees in nine study areas provided the data for Table 15.

The data reveal that the leading canopy species listed are not always leaders in reproduction. Some of the most shade tolerant species show the lowest reproduction as, for example, red maple. This seems to indicate some other factor than shade may be involved —perhaps winter-time floods or floods at another period. It has been noticed by Smith (1953) that poorly drained soils usually supported luxurious understories of shrubs and herbs, and cutting so stimulated those layers that they limited desirable tree reproduction.

A total of 7 small trees, 7 shrubs, and 4 vines were present among 37 species in the reproduction. Most important among the vines were *Campsis radicans*, *Vitis* spp., *Smilax* spp., and *Rhus radicans*. The most important shrub was *Cephalanthus occidentalis*, followed by *Ilex decidua* and *Lindera benzoin*. *Cercis canadensis* and *Asimina triloba* were the leading small tree species.

14. *Ranking of 20 most important species of herbaceous understory of 18 bottomland hardwood areas*

SPECIES	Constance	Abundance
Poison Ivy (Rhus radicans)	100	1
Aster (Aster ontarionis)	100	1
Violet (Viola spp.)	100	1
Rice Cut Grass (Leersia spp.)	94	3
Trumpet Creeper (Campsis radicans)	77	3
Sedge (Carex muskingumensis)	67	3
Fringed Loosestrife (Lysimachia ciliata)	61	2
Virginia Wild Rye (Elymus virginicus)	50	1
Smartweed (Polygonum virginianum)	50	3
Clearweed (Pilea pumila)	50	1
Green Dragon (Arisaema dracontium)	44	3
Virginia Creeper (Parthenocissus quinquefolia)	44	2
Cane (Arundinaria gigantea)	44	4
Coralberry (Symphoricarpos orbiculatus)	44	3
Grape (Vitis spp.)	44	2
Wood Nettle (Laportea canadensis)	33	4
Cat Brier (Smilax spp.)	28	2
Moonseed Vine (Menispermum canadense)	28	2
Lizard Tail (Saururus cernuus)	28	2
Sedge (Carex grayi)	28	2

Arranged in order of decreasing importance based upon the percentage of the areas in which occurred (constance) and arbitrary scale of abundance, 1—most abundant, 5—least abundant.

SPECIES	0–1"	1–2"	2–3"
Ostrya virginiana	8	1	0
Campsis radicans	12	0	0
Acer negundo	18	3	1
Vitis spp.	12	0	0
Ulmus americana *	15	21	3
Celtis occidentalis *	35	8	5
Ilex decidua	15	1	1
Fraxinus spp.	60	7	1
Euonymus atropurpureus	1	0	0
Cercis canadensis	15	5	0
Quercus macrocarpa	2	0	0
Carya spp.	48	3	2
Ulmus rubra	27	2	1
Smilax spp.	1	0	0
Cephalanthus occidentalis	38	0	0
Acer rubrum	8	0	1
Morus rubra	9	3	0
Sambucus canadensis	6	0	0
Acer saccharum	5	0	0
Juglans nigra	1	0	0
Quercus palustris	6	0	4
Gymnocladus diocia	3	0	0
Asimina triloba	12	0	0
Cornus racemosa	6	3	0
Crataegus spp.	7	1	0
Rhus radicans	9	0	0
Symphoricarpos orbiculatus	6	0	0
Quercus velutina	4	3	0
Quercus muhlenbergii	2	0	0
Forestiera acuminata	1	0	0
Liquidambar styraciflua	18	0	0
Carya illinoensis	5	1	2
Nyssa sylvatica	5	0	0
Sassafras albidum	8	0	0
Diospyros virginiana	2	0	0
Lindera benzoin	10	0	0
Quercus bicolor	2	1	0
Acer negundo	1	0	0
Quercus alba	3	0	0
Hamamelis virginiana	1	0	0
Maclura pomifera	1	0	0
Gleditsia triacanthos	0	1	0
Prunus serotina	2	0	0
Quercus imbricaria	1	0	0

15. *Diameter classes of shrub and tree reproduction, plants over one-foot-high, in nine lowland areas of lower Wabash Valley forest*

Data from 381 arm-length wide transects between randomly selected trees.
* One example 3"–4".

Betula nigra-Acer saccharinum/Pilea *Community*

A long stretch of southern Illinois river vegetation reveals the presence of several well-marked communities. Relative to their frequency of occurrence along the banks, they may be listed as black willow, silver maple, cottonwood, and river birch. River birch communities occur on poorly drained soil where organic matter is low. This is due to frequent inundations and currents of the stream which carry away practically all organic materials. Usually only piles of ungerminated maple fruits, too water-soaked to germinate, and driftwood are left by water subsidence. Aggrading influences are shown by distribution of organic matter and sand in the soil profile. Greater amounts of organic matter are frequently found several inches below the surface rather than following a gradual decline with depth as in other areas. Greater amounts of sand than are found in surface layers may also exist buried. Sand is generally quite low in the mechanical analysis, however.

Ten 100-square-meter quadrats were taken at random in a river birch stand where all trees were 4 to 10 inches DBH. Three species, making up 127 trees, were recorded. River birch constituted 81.9 per cent, silver maple 17.3 per cent, and green ash 0.8 per cent of the stand. Light on the forest floor was poor and a thin understory of *Pilea pumila, Geum canadense, Aster ontarionis, Belamcanda chinensis, Laportea canadensis,* and *Cinna arundinacea* occurred. These herbs formed rather unmixed groupings. Only seedlings of trees occurring were silver maple and ash, both of which were rare. In the lowest places there was virtually no understory.

In another birch community where trees were 4 to 8 inches DBH, ten 100-square-meter quadrats revealed 156 trees of which river birch constituted 62.3 per cent of the stand. Others included American elm 8.9 per cent, silver maple 14.7 per cent, red maple 0.2 per cent, and white ash 11.5 per cent. The trees were even-aged and shade was heavy. The forest floor was clean and there were no seedling trees. In some of the more open areas, an understory of *Leersia oryzoides, Impatiens biflora,* and *Polygonum hydropiperoides* occurred.

It has been shown in Cumberland County (Frits and Kirkland, 1960) that river birch is common along streams with more acid soils which drain areas of Illinoian age and is absent along streams carrying more alkaline materials out of the Wisconsin glacial moraine. It

Silver Maple (Acer saccharinum)

Silver maple, a widespread tree of the eastern deciduous forest, is also known as soft maple, river maple, water maple, and swamp maple. These names identify its position in the moisture gradient of the environment. It may frequently occur in nearly pure stands along floodplains of rivers where it also may grow with cottonwood and willow. It is not uncommon for silver maple to make up nearly half the composition where a limited number of other dominants occurs (Table 16). Its commonest associates are other pioneer species. Besides those already mentioned, it may occur with box elder, elms, and ash (Fig. 28). These are all intolerant trees, soon losing their canopy

16. *Percentage of frequency and composition of tree species over 4" DBH in an* Acer saccharinum-Populus deltoides/Aster *community near New Haven, Illinois*

SPECIES	Frequency	Composition
Acer saccharinum	25	29.0
Populus deltoides	2	10.3
Acer rubrum	2	9.3
Carya illinoensis	2	9.2
Liquidambar styraciflua	12	9.0
Ulmus americana	16	6.1
Populus heterophylla	6	5.7
Platanus occidentalis	6	5.1
Quercus palustris	6	5.1
Fraxinus lanceolata	5	4.2
Salix nigra	2	2.6
Betula nigra	4	2.1
Carya ovata	1	1.1
Juglans nigra	1	0.7
Gleditsia triacanthos	1	0.4
Quercus macrocarpa	1	0.1

Data from 50 random pairs, July 1959.

28. [left] A community of silver maple, elm, and ash in the lower Wabash valley near New Haven. [right] A community of silver maple and ash showing various flood-water marks on trees. The understory is principally lizard's tail (*Saururus cernuus*).

position in a shade-increasing environment of later successional stages. Commonest understory species are of *Pilea, Aster,* and lizard's-tail (Table 17).

Young branches are gray in color, with twigs being somewhat reddish. Numerous accessory buds appear in reddish clusters. Once the buds begin to swell, they usually open within a few days and the greenish-yellow or sometimes reddish apetalous flowers are revealed before the leaves begin to unfold. Flowering in our area very reliably takes place in February. Fruits are formed within a few weeks and are shed by wind in April and May when they are mature. Dispersal is usually local except in the case of water and distance migration by this agent is not known. Germination after 32 days soaking has been found by Hosner (1957) to be but 8 per cent less than controls which had 49 per cent germination.

When shade is not too great and litter is not deep, the seedlings develop rapidly. Competition between them is high as there are often as many as 200 to 300 seedlings per meter. Seedling survival after 31 days of near total submergence is good, though growth is much slowed and recovery likewise slow (Table 18).

Pin Oak (Quercus palustris)

Pin oak forms nearly pure stands (Tables 19 and 20) on heavy soils where drainage is poor (Minckler, 1957). In our area there are many acres of pin oak on level lowland areas. These "flats" are seasonally wet—mostly winter and spring. They tend to be dry in late summer and early fall.

The physiognomy of the pin oak community is distinct. The

17. *Percentage of composition, based on foliage cover, and frequency of an* Acer saccharinum-Populus deltoides/Aster *understory community near New Haven, Illinois*

SPECIES	Frequency	Composition
Pilea pumila	6	14.0
Aster spp.	7	12.8
Saururus cernuus	5	12.0
Viola spp.	5	6.7
Carex muskingumensis	4	6.5
Leersia spp.	3	5.4
Lindernia dubia	2	4.8
Acer saccharinum	2	4.6
Arundinaria gigantea	2	4.3
Campsis radicans	3	4.1
Uniola latifolia	2	2.9
Rhus radicans	2	2.9
Carex typhina	2	2.6
Asclepias perennis	1	2.4
Menispermum canadense	1	2.4
Ulmus seedlings	1	2.4
Laportea canadensis	2	1.7
Polygonum spp.	2	1.4
Ambrosia trifida	1	1.2
Lysimachia ciliata	1	0.8
Bidens spp.	1	0.7
Celastrus scandens	1	0.7
Oxalis cymosa	1	0.7
Quercus palustris	1	0.7
Acer rubrum seedlings	1	0.6
Gleditsia triacanthos	1	0.5

Data from 50 one-meter quadrats with stratified random distribution, July 1959.

strongly expressed excurrent form of pin oak and its usually even age in a stand make recognition of the type outstandingly easy. Hawthorn and deciduous holly are both rather common small tree constituents of the second layer. The understory usually contains such ubiquitous species as poison ivy (*Rhus radicans*), trumpet creeper (*Campsis radicans*), and aster (*A. simplex*). Presence of certain sedges, how-

18. *Survival and growth of silver maple seedlings submerged for 31 days*

SIZE AND WEIGHT	Control Group	Submerged Group
Average Height	15.5"	10.0"
Average Weight of Tops	25.1 gm.	9.8 gm.
Average Weight of Roots	7.7 gm.	2.3 gm.
Average Length of Roots	1008.4 cm.	199.0 cm.

The submerged group consisted of 28 specimens of which 23 survived, the control group of 24 specimens. Both groups had an average height of 9" at the beginning of the period.

19. *Percentage of frequency and composition of tree species over 4" DBH in a* Quercus palustris/Carex hyalinolepis *community, southwestern Jackson County*

SPECIES	Frequency	Composition
Quercus palustris	62	74.1
Carya ovata	6	7.3
Acer saccharinum	4	5.4
Ulmus americana	12	3.3
Quercus macrocarpa	5	3.2
Carya laciniosa	5	2.9
Liquidambar styraciflua	4	1.7
Carya illinoensis	1	0.6
Acer rubrum	1	0.5
Juglans nigra	1	0.4
Quercus bicolor	1	0.3
Fraxinus americana	1	0.1

Data from 50 random pairs.

20. *Percentage of frequency and composition of tree species over 4" DBH in a pin-oak community, Lawrence County*

SPECIES	Frequency	Composition
Quercus palustris	171	76.6
Quercus bicolor	14	9.0
Quercus lyrata	12	8.1
Quercus macrocarpa	5	3.6
Ulmus rubra	3	1.9
Quercus alba	1	0.3
Crataegus collina	2	0.2
Diospyros virginiana	2	0.2

Data from 105 random pairs.

ever, may be used as a site indicator for this type (Fig. 29). *Carex hyalinolepis* is often widely present through the understory of pin oak communities in both the Mississippi and Wabash lowlands of

THE LOWLAND SERIES 117

southern Illinois. Other sedges, such as *C. typhina* and *C. muskingumensis,* while not of such high frequency, are invariably found in the community type (Table 21).

The most common tree associates of pin oak are elms, ashes, sweetgums, and shumard and cherrybark oaks (*Quercus falcata* var. *pagodaefolia*). Pin oak seemingly occupies an intermediate position in the gradient of flood wetness and mature trees are less tolerant of standing water than most bottomland species. Mature trees were found to be killed by continuous flooding by the end of a two to three year period (Yeager, 1949). Flowering occurs in April or May, about the time the leaves appear. Both staminate and pistillate flowers are borne on the same tree. Acorns develop in 16 to 18 months, ripen, and fall in autumn, and germinate the next spring (Minckler, 1957).

29. Pin-oak community members: (1) pin oak (*Quercus palustris*), (2) *Boehmeria cylindrica*, (3) sedge (*Carex muskingumensis*), (4) sedge (*Carex hyalinolepis*), (5) *Ampelopsis cordata*.

21.

Percentage of composition, based on foliage cover, and frequency of a Quercus palustris/Carex hyalinolepis *community, southwestern Jackson County*

SPECIES	Frequency	Composition
Carex hyalinolepis	35	25.0
Rhus radicans	30	24.4
Campsis radicans	11	5.3
Carex muskingumensis	15	5.0
Carex spp.	10	4.5
Leersia lenticularis	12	3.8
Vitis spp.	13	3.8
Lysimachia ciliata	5	3.2
Parthenocissus quinquefolia	9	2.8
Impatiens biflora	3	2.1
Carya spp.	5	1.8
Apios americana	1	1.5
Menispermum canadense	3	1.5
Spermacoce glabra	3	1.5
Verbesina alternifolia	8	1.4
Viola spp.	8	1.2
Convolvulus spp.	4	1.2
Agastache nepetoides	1	1.1
Carex typhina	1	0.8
Elymus virginicus	3	0.8
Celtis occidentalis	2	0.7
Laportea canadensis	1	0.6
Fraxinus americana	3	0.5
Rubus spp.	2	0.5
Cassia marilandica	1	0.4
Gleditsia triacanthos	2	0.4
Mimulus alatus	1	0.4
Onoclea sensibilis	2	0.4
Ulmus spp.	4	0.4
Dioscorea villosa	2	0.3
Polygonum opelousanum	1	0.3
Quercus spp.	3	0.3
Acalypha spp.	1	0.2
Apocynum spp.	1	0.2
Boehmeria cylindrica	1	0.2
Potentilla simplex	1	0.2
Oxalis cymosa	1	0.2
Sanicula spp.	2	0.2
Polygonum virginianum	1	0.1
Stachys tenuifolia	2	0.1
Cryptotaenia canadensis	1	0.0
Panicum spp.	1	0.0
Pilea pumila	1	0.0

Data from 105 one-meter quadrats, September 1958.

Sycamore (Platanus occidentalis)

American sycamore, a handsome stalwart tree of bottomlands, is characterized by a light gray outer bark which scales away upon the upper extremities of branches and trunk to reveal an inner bark of chalky white to pale green color. This pallid color, by moonlight, uniquely marks the river's edge for navigators and has made the sycamore a riverman's favorite (Eifert, 1959).

Sycamore is a fast growing tree and becomes larger in diameter than any other American hardwood (Merz, 1958). It has been said to be to the deciduous forest what the redwood is to the coastal forest. Sargent (1933) states that a diameter of 11 feet and a height of 170 feet may be attained. Britton and Brown (1923) state that this species reaches a maximum height of about 130 feet and a trunk diameter of 14 feet.

Two sycamore trees in the Wabash valley within a few miles of each other and on opposite sides of the river have been reported for their outstanding size. These trees were perhaps the largest ever known to exist between the Allegheny and Rocky mountains. "A giant sycamore is monarch of the Wabash River bottoms in Gibson County (Ind.). Eleven feet in diameter, thirty-three feet in circumference, and 150 feet high, the lone sycamore stands in the middle of a small field overlooking the Wabash River at a point 2 miles southeast of Mt. Carmel. There is no such tree in Gibson County and nothing like it is known in this part of the state as far as is known" (Anonymous, 1921). Another sycamore across the river in Illinois "stood on the banks of Coffee Creek, a few hundred feet from where the creek empties into the great Wabash at Rochester and about four miles below Mt. Carmel in Wabash County" (Risely, 1911). This tree was nearly nine feet in diameter. It was cut down in 1897 by its owner who wished to avoid hordes of people who came to view it.

Some special blend of river silt fertility from floods and moisture relations as well as climatic favor must combine in the Wabash Valley to invigorate these hardwood denizens. Travelers and explorers are reported to have used large hollowed sycamores for shelter, being able, in fact, to ride into their basal hollows on horseback. Hollow sycamore logs of large size also have been used as well casings.

Unlike several of its lowland companions, the willow, cottonwood, and ash, sycamore is monoecious, which is to say both sexes are on the same tree. Staminate and pistillate flowers are borne separately in pendant greenish balls about an inch or more in diameter. The pistillate flowers produce seeds which are tightly packed around a central core to make a ball-like fruit. These seeds are vertically elongated and surrounded by brownish hairs which serve to buoy them on the wind currents. Flowers appear about April and the seed-ball, a multiple type fruit, ripens in fall. These often remain on the tree in winter and fall to earth in spring. Birds, the finches chiefly, may break them open and scatter the seeds. These achenes of the sycamore are light in weight, averaging about 200,000 per pound (Merz, 1958). Average germinative capacity has been reported by Toumey (1931) as being about 30 per cent. Owing to proximity of sycamore to rivers, a goodly amount of seed may settle upon water and subsequently be deposited on mud flats and wet shores where germination may occur.

Sycamore is favored by good light and, as a seedling, will grow rapidly. Its growth rate is exceeded only by some pines, cottonwood, silver maple, and willow (Merz, 1958). Sycamore displays an intermediate tolerance and a good competitive ability as it has been found

22. *Percentage of frequency and composition of tree species over 4″ DBH in an* Acer negundo-Platanus/Rhus radicans *community along Mississippi River*

SPECIES	Frequency	Composition
Acer negundo	34	31.4
Platanus occidentalis	34	30.2
Ulmus americana	14	12.1
Acer saccharinum	14	11.0
Fraxinus lanceolata	7	5.1
Populus deltoides	4	3.5
Salix nigra	4	2.9
Celtis laevigata	2	1.7
Morus rubra	2	1.2
Carya illinoensis	2	0.6
Maclura pomifera	2	0.2

Data from 50 random pairs one mile south of Route 146, Union County, August 1959.

23. *Percentage of frequency and composition of a box elder-sycamore community along Mississippi River*

SPECIES	Frequency	Composition
Acer saccharinum	8	11.0
Acer negundo	41	31.4
Populus deltoides	3	3.5
Ulmus americana	7	12.1
Fraxinus pennsylvanica	7	5.1
Salix nigra	2	2.9
Platanus occidentalis	24	30.2
Maclura pomifera	1	0.2
Celtis laevigata	1	1.7
Carya illinoensis	2	0.6
Morus rubra	4	1.2

Data from 50 random pairs, all trees over 4" DBH, Alexander County.

to be a member of every successional stage to and including climax. Its usual associates are other pioneer species such as silver maple, box elder, elms, ash, and black willow (Tables 22, 23).

The understory of communities dominated by sycamore consists of a number of ubiquitous species such as poison ivy, aster, trumpet creeper, and Virginia creeper. For this reason, as well as the fact that sycamore itself tends to enter into so many other communities, a mixed or transitional type often is manifested. This makes for a less distinctive community type. Other herbs often seen in the understory are *Ampelopsis arborea*, jewel weed, and *Boehmeria cylindrica* (Table 24).

Quercus-Carya/Hymenocallis *Community*

This community is one of the stages in the transition from a swampy situation to the oak-hickory forests of southern Illinois. As conditions become drier, this community gives rise to one which is again dominated by *Quercus* and *Carya*, but in which the species composition is drastically different, particularly in the herbaceous union.

Dominant in this community are the species which still require a considerable amount of water. *Quercus falcata* var. *pagodaefolia*

24. *Percentage of frequency and composition, understory of an* Acer negundo-Platanus/Rhus radicans *community along the Mississippi River*

SPECIES	Frequency	Composition
Rhus radicans	45	50.9
Aster spp.	13	10.3
Impatiens spp.	7	8.5
Parthenocissus quinquefolia	12	6.7
Campsis radicans	9	5.2
Sambucus canadensis	2	2.9
Acer negundo	5	2.4
Lonicera japonica	1	1.9
Vitis spp.	1	1.7
Viola spp.	4	1.6
Boehmeria cylindrica	1	1.5
Menispermum canadense	2	1.3
Sanicula spp.	2	1.3
Ampelopsis arborea	3	1.1
Eupatorium spp.	3	1.1
Gleditsia triacanthos	1	0.6
Celtis occidentalis	1	0.4
Sida spinosa	2	0.3
Forestiera acuminata	1	0.2
Rubus spp.	2	0.2

Data from 50 one-meter quadrats with stratified random distribution, one mile south of Route 146, Union County, August 1959.

occurs in great abundance, while *Quercus palustris, Q. michauxii, Carya ovata,* and *C. ovalis* are plentiful (Table 25). The presence of *Nyssa sylvatica* and *Liquidambar styraciflua* indicates this community to be one where moisture is abundant.

There is a scarcity of shrubs in this community due apparently to the rather close spacing of the trees. Those which do occur are mostly poorly developed, the chief ones being *Asimina triloba* and *Lindera benzoin.* The herbaceous layer is distinctive in this community, the most conspicuous member being *Hymenocallis occidentalis.* Other prominent species are *Habenaria peramoena, Spigelia marilandica,*

25. *Percentage of frequency and composition of tree species over 4" DBH in a* Quercus-Carya/Hymenocallis *community*

SPECIES	Frequency	Composition
Quercus falcata var. pagodaefolia	40	29.0
Quercus palustris	15	10.5
Carya ovata	15	10.5
Liquidambar styraciflua	12	7.9
Quercus velutina	12	7.9
Nyssa sylvatica	10	6.3
Carya ovalis	10	5.2
Quercus michauxii	7	5.2
Quercus lyrata	7	5.0
Quercus imbricaria	7	5.0
Quercus rubra	4	2.6
Quercus shumardii	4	2.6
Ulmus rubra	4	2.6

Data from 38 random pairs near Dongola in Union County.

and the grasses *Leersia virginica* and *Glyceria striata*. *Asclepias perennis* is rather common. Most of the species in the *Hymenocallis* union belong to the summer and fall aspects.

Fagus-Liquidambar/Rhus *Community*

Whenever deep swampy situations give way to mesic woodlands, the *Fagus-Liquidambar/Rhus* community becomes prevalent. It is a transition community between true swampy and mesic forest which eventually will be dominated by the *Fagus-Acer saccharum* association. It is not widespread in southern Illinois, but frequently occurs in areas where drainage is rather poor and where the canopy is so dense that intense shading occurs. The principal species of the tree layer are *Fagus grandifolia* and *Liquidambar styraciflua* which comprise 37.5 per cent of the woody species. Both of these trees attain considerable size with some *Fagus* specimens having a DBH of 40 inches. Subordinate species include the stately *Liriodendron tulipifera, Ulmus rubra, Acer rubrum,* and *Prunus serotina*. In all, 22 tree species have been recorded from the upper layer of this community

(Table 26). As the *Fagus-Liquidambar/Rhus* community becomes slightly less mesic, *Acer saccharum* begins to establish itself, gradually replacing *Liquidambar* as the co-dominant.

The dense shading and abundant moisture is conducive to an abundance of mesophytic shrubs. Woody vines such as *Rhus radicans* and species of *Vitis* abound. In considerable numbers are the sweet-smelling *Lindera benzoin*, the spiny *Aralia spinosa, Rhus glabra,* and *Asimina triloba*. These species are all a conspicuous part of the forthcoming *Fagus-Acer saccharum* community which follows. *Corylus americana* is frequent along the drier margins of the community.

26. *Percentage of frequency and composition of tree species over 4" DBH in a* Fagus-Liquidambar/Rhus *community, Union County*

SPECIES	Frequency	Composition
Fagus grandifolia	30	21.7
Liquidambar styraciflua	22	15.9
Liriodendron tulipifera	11	8.7
Ulmus rubra	8	5.8
Prunus serotina	8	5.8
Acer rubrum	8	5.2
Acer saccharum	8	4.3
Juglans nigra	8	4.3
Quercus velutina	4	2.9
Cornus florida	4	2.9
Carya glabra	4	2.9
Quercus alba	4	2.9
Diospyros virginiana	4	2.9
Carya laciniosa	4	2.6
Celtis occidentalis	2	1.4
Morus rubra	2	1.4
Cercis canadensis	2	1.4
Sassafras albidum	2	1.4
Carpinus caroliniana	2	1.4
Quercus muhlenbergii	2	1.4
Populus deltoides	2	1.4
Carya cordiformis	2	1.4

Data from 35 random pairs.

Foliage cover in this community is sparse, probably due to the great abundance and subsequent choking out of herbaceous species by the trailing and climbing *Rhus radicans*. The few herbaceous plants which do occur are mostly members of the prevernal and vernal aspects, apparently able to receive sufficient moisture from the soil before competition becomes too great. Only *Campanula americana* and *Eupatorium incarnatum* are met with any regularity during the late summer aspect.

The *Fagus-Liquidambar/Rhus* community is found in suitable situations throughout the southern seven counties. It may also be detected at a few local stations in extreme southern Jackson, Williamson, and Saline Counties.

Annotated List of Bottomland Areas

North Fork, Saline River	Sect. 5, T 7 S, R 7 E
New Haven	Sect. 21, T 7 S, R 10 E
Beaver Crossing, near Calvin	Sect. 27, T 7 S, R 10 E
Little Wabash, near Emma	Sect. 33, T 6 S, R 10 E
South of Mt. Carmel	Sect. 2, T 2 S, R 12 W
Calvin	Sect. 28, T 7 S, R 10 E
East of Inman	Sect. 12, T 8 S, R 9 E
Cattail Slough, S. of New Haven	Sect. 31, T 7 S, R 10 E
North of New Haven	Sect. 17, T 6 S, R 10 E

Ravines

As a ravine is formed by widening and deepening due to landslide action, lateral cutting, and side gullies, a considerable growth of vegetation develops. Ravine conditions are very favorable for plants. It is in these moist woodland ravines that the most luxurious vegetation of southern Illinois is found. It is here that forests of beech, sugar maple, tulip tree, sourgum, species of hickory, and hackberry occur. Characteristic shrubs are bladdernut and spicebush. Some rare shrubs also occur. These are *Viburnum lentago* and *Rhus typhina,* two species known elsewhere in Illinois only from the northern counties. The herbaceous plants are for the most part vernal. Some of the more common are violets, dutchman's breeches (*Dicentra cucullaria*), squirrel corn (*Dicentra canadensis*), trilliums (*Trillium sessile* and *T. recurvatum*), bloodroot (*Sanguinaria canadensis*), rue anemone (*Anemonella thalictroides*) (Fig. 30), and harbinger-of-spring

30. [left] Rue anemone (*Anemonella thalictroides*) is a common species in a myriad of spring blooming herbs of ravine forest floors. [right] Bloodroot (*Sanguinaria canadensis*) is a familiar early blooming species of the ravine forest understory.

(*Erigenia bulbosa*). Rarer species such as blue-eyed mary (*Collinsia verna*), *Viola cucullata*, hepatica (*Hepatica triloba*), *Valeriana pauciflora* (Fig. 31), and various orchids (*Cypripedium parviflorum, Aplectrum hyemale, Corallorhiza wisteriana,* and *Orchis spectabilis*) occur in this habitat (Fig. 32). A notable fall-flowering orchid, *Triphora trianthophora*, may be found in the moist ravine at Little Grand Canyon (Jackson County). Many ferns also occur in the ravine woods. These are the areas which have a rapid development into excellent forest lands because of the high collection and preservation of moisture.

Rocky ravines are common throughout southern Illinois. Here precipitous slopes and bluffs occur and lateral cutting is slower. Rocky gorges are occasionally formed and these harbor a large number of mosses and liverworts on their dripping vertical walls. While most of the species of the clay ravines also occur in the sandstone rocky ravines, many additional ones find their occurrence here. The more exciting ones to find are clubmoss (*Lycopodium lucidulum* var. *occidentale*), maidenhair spleenwort (*Asplenium trichomanes*), bishop's cap (*Mitella diphylla*), *Saxifraga forbesii, Carex careyana,* wild leek (*Allium tricoccum*), trout lily (*Erythronium americanum* and *E. albidum*), rattlesnake orchid (*Goodyera pubescens*), *Aralia racemosa,* French's shooting star (*Dodecatheon frenchii*), *Synandra hispidula,*

31. [left] Hepatica (*Hepatica acutiloba*) on a rocky ledge in a ravine forest at Jackson Hollow, Pope County. [right] Valerian (*Valeriana pauciflora*) against a background of marginal fern.

and partridge berry (*Mitchella repens*). The rocky ravines possess a great number of rare southern Illinois species.

Springs

Precipitation penetrating porous surface and subsurface materials under the influence of gravity may ultimately come in contact with an impervious layer and seek an outlet along it at a lower level. This is the usual manifestation of springs and in our area the surface and substrate materials are either sandstone or, less commonly, limestone. The spring outlets are usually situated at the base of a bluff and would have to be considered small in size. These springs may also commonly occur in valley-sides where stream-cutting has deepened ravines to expose a water-bearing rock. Spring flow in southern Illinois is never great in the manner of the larger springs in Missouri. The escape of water here is a tranquil, almost unobservable flow. Only one spring observed in this study showed any escape of underground water with a bubbling or boiling of the sandy substrate. Coulter Spring in Grindstaff Hollow showed some bubbling or rolling of sand from a small fissure of the rocky and sandy bottom of a spring no more than sixteen square feet in area. A number of our small springs are classed as mineral springs because of the presence of

32. Plants of wooded ravines: (1) puttyroot orchid (*Aplectrum hyemale*), (2) delphinium (*Delphinium tricorne*), (3) yellow ladies slipper orchid (*Cypripedium calceolus*), (4) large white-flowered trillium (*Trillium gleasonii*), (5) French's shooting star (*Dodecatheon frenchii*).

sulphate of lime, carbonate of soda, aluminate of magnesium, sulphate of iron, and chloride of sodium.

Once a large spring was located at Wetaug. This was a limestone spring with a funnel-form basin some 30 feet in diameter. In earlier years its water was described as blue with a whitish hue, and reminiscent of the limestone springs of Missouri. Its discharge was considerable, forming a fine brook. It was suspected of being connected with the sink-hole ponds and caves in the St. Louis limestone of this area. The flow of this spring became muddy following one of the earth tremors the area has experienced and soon its flow declined and at present there is a saucer-shaped basin with a small sump in the center containing approximately a bucket of water. The spot is marked by a rank growth of giant ragweed and a box elder tree growing at its margin (Fig. 33).

One of the largest active springs in the area is located at the base of a limestone bluff in the LaRue Scenic Area. This spring issues from beneath the bluff with a steady flow of water very uniform in temperature. It is from these waters that blind cave fish have been taken.

One noted salt spring may be mentioned for its source of salt to early settlers in the area. A springy area in Saline County near Equality was once the scene of a thriving salt industry. The area is still littered with ruins of this activity. The waters of the spring, located near to the banks of the Saline River, test at four per cent. One plant with a notable tolerance to weak concentrations of salt is found here. It is the shadscale (*Atriplex hastata*).

Because of generally lower temperature of spring water from other aquatic habitats, it was thought that a vascular flora somewhat different would result. An inseparable association of spring branches and alder growth in southern Illinois has been observed. The water cress (*Nasturtium aquatica*) invariably indicates water of a cooler nature as does the presence of *Scirpus validus, Glyceria pallida, Galium tinctorium, Elodea canadensis, Callitriche heterophylla, Chelone glabra, Polygonum sagittatum, Glyceria septentrionalis,* and

33. [left] Formerly a big spring of bluish-white water existed near Wetaug. Its location is marked by the box elder tree in the center of the picture. [right] One of numerous small springs at the base of limestone bluffs at Pine Hills, Union County.

Alopecurus aequalis (Fig. 34). These plants are mainly those with northern affinities.

On cherty areas around the spring or on banks near to the source and where water may sometimes run over the lower parts of plants may be found the brookweed (*Samolus parviflorus*), bedstraw (*Galium triflorum*), *Cardamine arenicola, Polygonum punctatum,* and *Cornus racemosa.* In rocky shallow stream beds of the spring branches frequently are found *Peplis diandra* and *Dianthera americana.* On sandy banks of the spring branch some of the usual plants are: *Juncus secundus, Juncus nodatus, Cyperus aristatus, Mentha canadensis, Ranunculus micranthus,* and *Senecio aureus.*

Away from the source or orifice of springs, the trickles or brooks are known as spring branches. Along the banks of spring branches before they unite with other streams may be found *Mentha canadensis, Polygonum setaceum, Polygonum punctatum, Scutellaria lateriflora,*

34. Plants of fresh-water springs: (1) tear thumb (*Polygonum sagittatum*), (2) waterweed (*Anacharis occidentalis*), (3) water starwort (*Callitriche heterophylla*), (4) water cress (*Nasturtium officinale*), (5) bulrush (*Scirpus validus*).

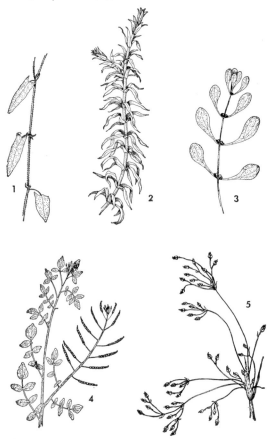

Carex frankii, Carex crinita, Scrophularia marilandica, Teucrium canadense, and *Senecio aureus.*

Annotated List of Fresh or Salt Water Springs

Clear Springs	Sect. 27, T 10 S, R 3 W	Jackson Co.
Midland Hills	Sect. 18, T 10 S, R 1 W	Jackson Co.
Bluff Lake	Sect. 17, T 13 S, R 3 W	Union Co.
Seminary Fork	Sect. 22, T 11 S, R 2 W	Union Co.
Pine Hills	Sect. 1, T 11 S, R 2 W	Union Co.
	Sect. 9, T 11 S, R 3 W	Union Co.
Dixon Springs	Sect. 16, T 13 S, R 5 E	Pope Co.
* Nigger Spring	Sect. 26, T 9 S, R 8 E	Gallatin Co.
Coulter Spring	Sect. 22, T 10 S, R 8 E	Pope Co.
Belle Smith Spring	Sect. 33, T 11 S, R 5 E	Pope Co.
Big Spring (Wetaug)	Sect. 2, T 13 S, R 1 W	Pulaski Co.
Ullin	Sect. 33, T 14, R 1 W	

* Salt spring near Equality

Seepages

Seepage areas may be of two kinds: rocky capstones on rock ledges of bluffs and the moist springy hillside as found in pastures and meadows. Though neither of these is usually thought of as a spring, they are formed in a similar manner. Plants occupying such situations are usually somewhat different from other habitats and each differs from the other.

The springy hillside seepage presents several species of rushes, sedges, and grasses. The following species are somewhat characteristic: *Cyperus flavescens, Eleocharis obtusa, Juncus tenuis,* and *Carex frankii.* Rocky seepage areas provide a pathway for water over a cliff and provide moisture to an otherwise xerophytic environment. Due to their ability to withstand washing, lichens may be the most abundant plants in this habitat. When depressions occur in the rock, organic materials will have a better chance to accumulate, and "miniature swamps" of temporary duration may be formed. A close investigation of the vegetation here may reveal quillwort (*Isoetes butleri*), a rare fern ally. Other species found in this habitat are rush (*Juncus tenuis*), false garlic (*Nothoscordum bivalve*), onion (*Allium canadense*), false dandelion (*Krigia dandelion*), *Plantago pusilla, Draba brachycarpa, Houstonia pusilla, Sagina decumbens, Callitriche terrestris,* and *Poa annua.*

A seepage area north of Murphysboro harbors several interesting plants. A small spring flows onto low ground where the drainage is poor, and this results in a marshy condition throughout the year. The area is very spongy underfoot and, until recent efforts to drain it were made, it actually was somewhat bog-like. From the immediate area of the orifice may be found water starwort (*Callitriche heterophylla*), while along the course of its flow to lower ground a dense stand of tear thumb (*Polygonum sagittatum*) is found. Where the waters first slow and spread over the land may be found the marsh fern (*Dryopteris thelypteris*) and turtle head (*Chelone glabra*). Farther away from the source where the land is lower and where several inches of water depth may prevail much of the time is found a group of sedges (*Carex comosa, C. crinita, C. lanuginosa,* and *C. tribuloides*). In deepest water near a roadside ditch are the swamp rose (*Rosa palustris*), cattail (*Typha latifolia*), wire wool (*Scirpus cyperinus*), and arrowhead (*Sagittaria latifolia*). Other plants scattered over the area include *Eupatorium perfoliatum, Iris virginica* var. *shrevei, Ranunculus pusillus, R. sceleratus, Solidago patula,* and arrow-wood (*Viburnum recognitum*).

Wet Roadside Ditch Community

This community presents an interesting array of species. Because of its nearness to highways, it is continuously disturbed. Mowing occurs in most of the ditches at one time or another during the growing season. Distribution of vegetation depends upon the degree of disturbance and the amount of water which remains in the ditches.

Certain areas of exceedingly poor drainage may remain inundated for most of the year. Usually, however, some ditches are wet throughout the winter and spring, but begin to dry up during early summer. In years of average rainfall, some ditches are very dry by late July and remain this way until October, except for brief periods following severe summer thunderstorms. Seedlings of swamp pin oak, cottonwood, and black willow may occur in these ditches, but trees are mostly nonexistent, due in part to periodic mowing.

Areas which are inundated throughout the year are scarce, and seldom occupy more than a few hundred square feet. Few species occur, although those which are able to tolerate standing water do occur in considerable numbers. Water starwort and the duckweeds (*Lemna minor* and *Spirodela polyrhiza*) are the common ones.

The remaining ditch areas are subject to varying lengths of inundation periods. Conspicuous colonies of perennial grasses, sedges, and rushes allow for few bare ground areas. Most grasses of these ditches are of a weedy nature and include Kentucky blue grass, goose grass (*Eleusine indica*), and finger grass (*Digitaria sanguinalis*). Other grasses of local abundance include species of *Leersia, Echinochloa, Paspalum*, and *Panicum agrostoides*. Sedges and rushes grow in areas more moist than that which supports these grasses. The spike-rush (*Eleocharis*) is well-represented by several species—the slender *E. acicularis*, the medium-sized *E. obtusa* and *E. tenuis* var. *verrucosa*, and the coarse *E. palustris*. The latter may occur in dense stands occupying several hundred square feet. A great variety of herbaceous species grows in the wet ditches, most of them occurring as scattered specimens. A few grow thickly unto themselves as, for example, *Polygonum setaceum* var. *interjectum* and *Isoetes melanopoda*.

Relatively few species flower in early spring. Some of the first are speedwell (*Veronica arvense*), small buttercup (*Ranunculus pusillus*), Missouri violet (*Viola missouriensis*), and bulbous cress (*Cardamine bulbosa*). Later conspicuous species are *Ammannia coccinea, Rotala ramosior, Gratiola neglecta, Gratiola virginica, Spermacoce glabra*, and numerous species of *Polygonum*.

The Upland Series

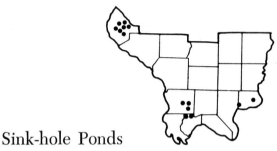

Sink-hole Ponds

Natural upland ponds in Missouri, relicts of a former period of leveling when the land was a wet, low-lying peneplain, have been found to harbor aquatic species formerly widespread in Tertiary time (Steyermark, 1951). Presence of numerous sink-holes of pre-glacial origin in southern Illinois leads to expectation of floristic similarities.

Sink-hole ponds generally are regarded as the result of water moving through underground limestone to dissolve it away and form an underground cavern. Subsequently, roofs of these caverns settle and cause surface depressions which become filled from drainage of surrounding land (Eddy, 1931). Others, generally smaller, seem not to be formed in this manner but from washing of loose, fine, underlying sand. These are usually dry and occur close to the head of abruptly descending ravines. A more resistant barrier between the sink and the ravine usually prevents the sink from uniting with the break and becoming its head (Engelmann, 1868).

Sink-hole ponds vary in depth. Almost all are round or oval as seen from the air (Fig. 35). Some are saucer-shaped and others are more funnel-shaped and deeper. Drainage may occur through this cone and thus provide a source for underground streams. One such stream disappears into the ground in Hardin County. It is called "Lost Creek" and is located a few hundred feet southwest of the Golconda

"Y," about 100 feet south of a concrete culvert (T 12 S, R 7 E, Sect. 23.) (Bonnell, 1946).

Drainage through the bottom cone may occur unexpectedly and cause an emptying of the entire sink (Fig. 36). This was observed to have happened during July of 1957 in a pond located just east of Highway 3, near Ellis Grove, in Randolph County (Bollwinkel, 1958).

In the Randolph-Monroe Counties area, sink-hole ponds have suffered heavy disturbance. Many ponds are used for stock-watering places, and consequently the banks have been made bare by trampling of pigs, cattle, and other animals. Many sink-holes of the more shallow type are drained and the drainage outlet maintained by placing coarse rocks or a perforated can over the hole to prevent its clogging (Fig. 36). In this case the shallow bottom and sides of the sink may be plowed, harrowed, and planted. In these farming enterprises the destruction of the flora has been all but complete.

The only possibility of studying and obtaining floristic information from sink-hole ponds, then, comes from a study of those which are still wooded and relatively undisturbed. Needless to say these are difficult to locate. An early study of sink-holes in southwestern Illinois was made by Hus (1908). His description of their vegetation at that time is worth noting. "Sinkholes in the deciduous forest may be either dry or wet. Where dry, i.e., where a drainage system is present, they harbor plants which are usually referred to as appreciating a well-drained soil. *Hydrastis canadensis, Oxalis violacea, Polygonatum giganteum* [*Polygonatum canaliculatum*], *Smilacina racemosa, Arisaema dracontium,* and *A. tryphyllum* occur in greatest abundance. They form the most important portion of the flora of sinkholes during the spring and summer months, together with *Sanicula marylandica*,

35. Land surface in Monroe and northern Randolph Counties pocked with small limestone sinks. (Photo by Southern Illinois University Photographic Service from an area near Renault.)

36. [left] A sudden drainage of a sink-hole pond near Ellis Grove in Randolph County shows a saucer-shaped depression and bottom cone. (Photo by Carl Bollwinkel.) [right] Device for keeping sink-hole open for drainage. The bottom of this shallow sink in Randolph County was under cultivation.

Spiraea aruncus [*Aruncus dioicus*], and occasional ferns as *Adiantum pedatum*, *Aspidium acrostichoides* [*Polystichum acrostichoides*], and *Cystopteris fragilis*. It is to be noted that *Hydrastis canadensis* forms large patches to the exclusion of all other species. *Oxalis violacea* ordinarily occurs singly. Occasionally *Passiflora lutea* is met with. The aerial portion of most of the plants just mentioned dies off in early fall, leaving the sinkhole comparatively bare. This forms a great contrast with the sinkholes occurring in the open plain. Here again dry and wet sinkholes are to be considered. The dry sinkhole in the plain is usually surrounded by shrubbery, composed chiefly of *Euonymus atropurpureus*, *Staphylea trifolia*, *Rhus toxicodendron* [*Rhus radicans*], *Pyrus coronaria*, *Sambucus canadensis*, and *Ribes gracile*, through which *Vitis cordifolia* [*V. vulpina*] twines. Occasionally a few oaks or hickories or a solitary *Platanus occidentalis* overtops the shrubbery, evidently a survival from a period when the drainage of the sinkhole was temporarily interrupted. The inner, upper portion of the sinkhole usually accommodates numerous smaller stunted specimens of *Ribes gracile* together with briars and brambles. The central portion of the sinkhole is ordinarily bare, the water, which at each rain, rushes past to escape through the fissure, sometimes as much as two feet in diameter, preventing the establishment of any plants. On the sides are found specimens of *Arisaema dracontium*, *Arisaema triphyllum*, and *Smilacina racemosa*. Neither *Hydrastis canadensis* nor *Commelina virginica*, so frequent in sinkholes in woods, is ordinarily met with here. Nor are the ferns which frequent the latter encountered. On the other hand, *Passiflora lutea* is far more common. When, as frequently happens, one of the edges of a sinkhole in the open plain is bare of shrubbery, *Hypericum drummondii, Desmodium dillenii,* and *Eryngium yuccaefolium* are found during the summer months.

"The wet sinkhole in woods, more familiar as a wooded pond, has a flora distinctly its own.

"Here and there the forest formation is interrupted because of small ponds which are the result of the obstruction of the drainage of sinkholes. As a result, not only is the sinkhole filled with water but the latter also accumulates over the ground which immediately surrounds the sinkhole and which ordinarily has a slight slope towards it. It is evident that the water will continue to encroach upon the ground till an equilibrium shall have been established between the water supply and the evaporation. Ordinarily the wooded ponds are therefore quite shallow with the exception of an area of great depth at or near the center. Hence we never find the central portion occupied by plants whose roots penetrate the mud, but by floating forms only.

"Wooded ponds permit the growth of more water-loving species of oak, occasionally associated with specimens of *Platanus occidentalis.* The larger amount of light permits the existence of numerous shrubs which, however, belong to comparatively few species, *Staphylea trifolia* being prominent at some distance from the water, where *Cephalanthus occidentalis* is abundant. *Vitis cordifolia* [*V. vulpina*], *Vitis riparia, Rhus toxicodendron* [*Rhus radicans*], *Smilax herbacea* [*S. lasioneuron*] and *S. hispida* are the principal creepers.

"The herbaceous flora is a distinct rosette formation, the xerophytic character possibly being due to too great an amount of water in the soil, interfering with the necessary life processes. . . . The principal components are *Arctium lappa minus, Solidago drummondii, S. canadensis, S. ulmifolia, Erigeron canadensis,* and *Prunella vulgaris.* As to the hydrophytic flora, *Heteranthera limosa* and *Pontederia cordata,* together with *Jussiaea repens* on the one hand, and *Lemna perpusilla* and *Wolffia punctata* on the other, are the principal species. The presence of large quantities of *Riccia natans* must also be mentioned.

"In sharp contrast to the wooded pond stands is the open pond, *i.e.*, the wet sinkhole in the field. All woody plants are absent with the exception of an occasional specimen of *Cephalanthus occidentalis*. The main flora around the wet sinkholes consists of grasses and sedges, chiefly *Cyperus acuminatus* and *C. erythrorhizos, Dulichium spathaceum* [*Dulichium arundinacium*], *Eleocharis acicularis, Scirpus lacustris* [*S. validus*], *Glyceria nervata* [*Glyceria striata*], and *Leersia oryzoides*. Ordinarily the wet sinkholes in the field are of small depth, the soil from the surrounding slopes being washed down in great quantities, not being retained by roots, etc., as in the case of a wooded pond. Hence the majority are sufficiently shallow to accommodate species of *Sagittaria*. Algae occur here in large quantities as they do in the wooded ponds or in the bodies of water in the American bottoms." (Hus, 1908).

The most common woody species surrounding limestone sinks in southern Illinois at present are *Fraxinus americana, Liquidambar styraciflua, Acer negundo, Ulmus rubra, Quercus palustris, Salix nigra, Juglans nigra,* and *Cephalanthus occidentalis* var. *pubescens*. The banks are frequently covered with brambles (*Rubus* spp.) and tall grasses such as stout wood reed (*Cinna arundinacea*). Submerged species are few, but *Najas flexilis, Utricularia gibba,* and *Myriophyllum heterophyllum* are found in some locations. Free floating forms are more common, but not all species of the water-meals occur in any one location of sink-hole pond. From all ponds examined, four species of watermeals were noted: *Wolffiella floridana, Wolffia columbiana, W. papulifera,* and *Spirodela polyrhiza*.

Attached floating forms are notably scarce, but include *Potamogeton diversifolius* and *P. pulcher*. Amphibious and wet meadow forms are more plentiful, with wet meadow plants in particular being well represented. The more uncommon wet meadow plants in sink-hole pond areas include *Ranunculus pusillus, R. oblongifolius, Callitriche heterophylla, Sagittaria graminea, Eleocharis acicularis, Echinodorus tenellus,* and *Heteranthera limosa*.

Decodon verticillatus and *Carex decomposita*, known from sink-hole ponds in Missouri, are found in southern Illinois only in swamps. Most of the plants unique to Missouri sink-holes have been found to be rare in southern Illinois and of 36 plants reported from this habitat in Missouri, only six are known from the same habitat in southern Illinois (Bollwinkel, 1958). One of these is *Carex brachyglossa* from Walker Pond, near Grand Tower.

Annotated List of Sink-hole Ponds

Near Burksville	T 3 S, R 10 W	Monroe Co.
East of Fults	Sect. 10, T 5 S, R 9 W	Monroe Co.
West of Ames	T 4 S, R 9 W	Monroe Co.
North of Renault	T 4 S, R 10 W	Monroe Co.
Northwest of Cave-in-Rock	Sect. 9, T 12 S, R 1 W	Hardin Co.
Near Cave-in-Rock	Sect. 3, T 12 S, R 9 E	Hardin Co.
Near Prairie du Rocher	T 5 S, R 9 W	Randolph Co.
West of Wetaug	Sect. 2, T 13 S, R 1 W	Pulaski Co.
Morgan Pond	Sect. 28, T 13 S, R 1 W	Union Co.

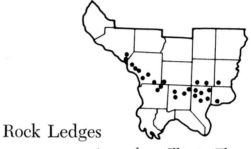

Rock Ledges

Two basic types of escarpments occur in southern Illinois. These are sandstone and limestone ledges (Fig. 37). Sandstone escarpments are found in an east-west direction across southern Illinois. Well-developed rock-ledge areas may be found at Rock Castle Creek in Randolph County; Pomona, Saltpeter Cave, and Giant City State Park in Jackson County; Wing Bluff and Panther's Den

37. Two basic types of escarpments in southern Illinois are sandstone (left), at Ferne Clyffe State Park, Johnson County, and limestone (right), at Pine Hills, Union County.

in Union County; Ferne Clyffe State Park in Johnson County; Belle Smith Springs, Burden Falls, and Jackson Hollow in Pope County. There are also areas in the Cave Hill, Gold Hill, and Wildcat Hill regions which extend northeast from the main body of the Shawnee Hills. Limestone escarpments occur chiefly along rivers—the Mississippi to the West and the Ohio to the East.

The sandstone escarpment usually supports a scrubby, xerophytic forest near its summit. The dominants here are post and black jack oaks, red cedar (*Juniperus virginiana*), and to a lesser extent winged elm (*Ulmus alata*). Farkleberry (*Vaccinium arboreum*) occurs as a characteristic shrub in this tree community. On gradual slopes, oaks and hickories are the characteristic canopy species. On sheer bluffs, little vegetation occurs except for a few vines which cling tenaciously to nearly barren walls. A nearly horizontal bare-rock shelf, at most only a few feet wide, is present near the top of many sandstone bluffs of the Shawneetown Ridge. In some localities, cliff-tops are broadly rounded exposed outcrops nearly 100 feet wide.

Winterringer and Vestal (1956) have discussed successional patterns which prevail on these sandstone ledges. Lichens, predominantly *Parmelia conspersa*, prevail on ledges and rock-slopes of the most fully exposed sites. On moderately exposed ledges and rock-slopes, the lichens are joined by two mosses (*Grimmia* and *Hedwigia*) and rarely the small clubmoss or rock selaginella (*Selaginella rupestris*). This latter species has been found on ledges overlooking Burden Falls and also at the summit of Jackson Hollow. On the least exposed ledges, although more shade prevails, there is still little vegetation except the typical lichens and mosses. In this situation, however, mosses are more abundant.

Thin soil areas occur as scattered patches and also as the narrow outer margin of soil borders. These areas support a vegetation of numerous lichens (*Parmelia* spp.), mosses (*Grimmia, Leucobryum,* and *Bryum*), and the tiny, succulent flowering plants of rock cress (*Sedum pulchellum*) and flower-of-an-hour (*Talinum parviflorum*) (Fig. 38). Larger succulent flowering plants such as cactus (*Opuntia humifusa*) and American aloe (*Agave virginica*) also occur.

Soil borders form the transition zone between the barren rock-ledges and the summit forests. Numerous grasses and other flowering plants occur, as well as an abundance of lichens and mosses. Notable grasses include poverty oat grass (*Danthonia spicata*), six-weeks

38. [top] Rock ledge at Indian Kitchen on Lusk Creek, Pope County, and [bottom] flower-of-an-hour which grows on sandstone ledges.

fescue (*Festuca octoflora*), and *Aristida* spp. Admixtures of little bluestem (*Andropogon scoparius*) and broomsedge (*A. virginicus*) occur as an understory to the canopy species in more open situations. Other frequently encountered plants of thin soil areas are false wild garlic (*Nothoscordum bivalve*), cleft phlox (*Phlox bifida*), species of bluets (*Houstonia* spp.), pinweeds (*Lechea* spp.), pussy-toes (*Antennaria* spp.), dittany (*Cunila origanoides*), pencil-flower (*Stylosanthes biflora*), *Psoralea psoralioides*, and *Panicum xalapense*. In shadier and consequently moister soil borders, species of *Heuchera* may play a conspicuous role. (For examples of plants of rock ledges, see Fig. 39.)

Crevices and recesses are ideal habitats for still other species. In exposed areas, *Sedum pulchellum*, *Opuntia humifusa*, St. Andrew's cross (*Ascyrum multicaule*), and woolly lip-fern (*Cheilanthes lanosa*) thrive (Fig. 40). In shadier crevices and recesses, woods fern (*Woodsia obtusa*) and wild columbine (*Aquilegia canadensis*) seem to do best.

39. Plants of rock ledges: (1) rock cress (*Sedum pulchellum*), (2) farkleberry (*Vaccinium arboreum*), (3) pencil flower (*Stylosanthes biflora*), (4) woolly-lip fern (*Cheilanthes lanosa*), (5) winged elm (*Ulmus alata*).

Drip-ways are described as being "wet streaks on gently sloping ledges, or narrow and flat outlets of small basins in the rock shelf" (Winterringer and Vestal, 1956). These drip-ways support several species of algae which make footing treacherous. With the algae may occur several kinds of lichens (*Peltigera, Dermatocarpon*, etc.) and mosses (*Grimmia, Polytrichum,* etc.). Frequent shallow depressions occur in sandstone rock. In these depressions, which hold water after spring rains, may be an accumulation of silt. Found here are Butler's quillwort (*Isoetes butleri*), sedges (*Cyperus aristatus*—Fig. 41—and *Bulbostylis capillaris*), small croton (*Crotonopsis elliptica*), false dandelions (*Krigia biflora* and *K. dandelion*), and others.

The number of Angiosperms which occurs along sandstone rock-ledges is remarkable. In 16 stations studied extensively by Winterringer and Vestal, there were recorded 145 species representing 111 genera and 67 families. High numbers of genera and families seem to be common in areas which are rather unfavorable to plant growth. Bluff-top vegetation is extremely sensitive to moisture fluctuations

THE UPLAND SERIES 143

which are brought about by periodic heavy rainfalls. In areas of little soil and scanty plant cover, heavy rains have tremendous eroding effects. Because of frequent summer droughts and exceptionally high surface rock temperatures in summer, most rock-ledge species flower early and die down by mid-summer. Bulter's quillwort dies down so soon that by June there is no trace of aerial portions of the plant.

Besides Butler's quillwort and rock selaginella mentioned previously, other rare species occur along sandstone rock-ledges. Black chokeberry (*Aronia melanocarpa*) has its only southern Illinois occurrence along the exposed bluff's edge of Old Stone Face in Saline County. The same locality is the only southern Illinois site for Mead's milkweed (*Asclepias meadii*). A sandstone bluff-top at Giant City State Park harbors the tiny white-fruited sedge (*Scleria nitida*).

40. [left] Woolly-lip fern (*Cheilanthes lanosa*), a xerophytic fern of rock ledges. [right] *Cyperus aristatus*, a small sedge inhabiting the thin mineral soil which accumulates in slight depressions of rock ledges.

41. [left] Saxifrage (*Saxifraga forbesii*) on a shaded, moist wall at Giant City State Park, Jackson County. [right] Carolina buckthorn (*Rhamnus caroliniana*), a shrub of southeastern affinity.

Moist Rock Walls

Rocks which are shaded by bluffs or forest trees and often moist from spring seepage provide a habitat for many attractive and rare plants. Liverworts, mosses, and many types of ferns can be observed growing on rocks and in crevices, utilizing the thin layer of organic material rendered available to them by their predecessors. Ferns typical of this kind of environment are polypody (*Polypodium virginianum*), walking-fern (*Camptosorus rhizophyllus*), and hay-scented fern (*Dennstaedtia punctilobula*). At wet edges of capstones in shaded places occur cinnamon fern (*Osmunda cinnamomea*) and regal fern (*O. regalis*). Clubmoss (*Lycopodium lucidulum* and *L. complanatum*) also occur on north-facing moist rockwalls or ledges. Two flowering plants growing in this habitat are bishop's cap and partridge-berry.

Cave-like Overhanging Rocks

The forces of erosion, primarily that of water, have produced overhangs of various sizes in the sandstone walls. Some of these caves are wet; others are dry. The moist-wall type of overhanging rock offers the most suitable habitat for French's shooting star (*Dodecatheon frenchii*). Although it is quite abundant in this habitat in a few places in southern Illinois, it is not found elsewhere.

On ledges or cave walls above French's shooting star, ferns of various types may be found intermingled with such flowering plants as alumroot (*Heuchera parviflora* var. *rugelii*), and saxifrage (*Saxifraga forbesii*) (Fig. 41). Below on the moist ground adjacent to the French's shooting star may be found waterleaf (*Hydrophyllum* spp.), clearweed (*Pilea pumila*), and *Cardamine pennslyvanica*. Under dry overhanging rocks the most common species is pellitory (*Parietaria pennsylvanica*) of the nettle family.

Annotated List of Sandstone Bluffs and Ravines

Beartrack Hollow	Pope Co.	Sects. 7, 17, 18, T 11 S, R 6 E
Blind Hollow	Hardin Co.	Sect. 18, T 11 S, R 10 E
Devil's Kitchen	Williamson Co.	Sects. 16, 21, T 10 S, R 1 E
Dixon Springs	Pope Co.	Sects. 28, 29, T 13 S, R 5 W
Dry Hill	Jackson Co.	Sect. 21, T 8 S, R 4 W
Ferne Clyffe	Johnson Co.	Sects. 22, 27, T 11 S, R 2 E
Fountain Bluff	Jackson Co.	Sects. 31, 36, T 9 S, R 3 W

THE UPLAND SERIES 145

		Sects. 1, 6, 7, 12, 13, 18, T 10 S, R 3 W
Giant City	Jackson and Union Cos.	Sects. 27, 28, 33, 34, T 10 S, R 1 W
Grindstaff Hollow	Pope Co.	Sects. 21, 22, 27, 28, T 10 S, R 8 E
Hooven Hollow	Hardin Co.	Sect. 22, T 11 S, R 9 E
Little Grand Canyon	Jackson Co.	Sect. 1, T 10 S, R 3 W
Jackson Hollow	Pope Co.	Sect. 18, T 11 S, R 5 E
Midland Hills	Jackson Co.	Sect. 18, T 10 S, R 1 W
Piney Creek	Randolph Co.	Sect. 23, T 7 S, R 5 W
Pomona Natural Bridge	Jackson Co.	Sect. 17, T 10 S, R 2 W
Pounds Hollow	Hardin Co.	Sects. 25, 35, 36, T 10 S, R 8 E

The River Bluff

Due to deepening and widening of valleys, there is an increase in exposure to wind, sunlight, and changes of temperature and a decrease in moisture content of the slopes. Mesophytic species are gradually replaced by the more xerophytic ones. Many of the river bluffs in southern Illinois are limestone, although there is no scarcity of sandstone.

Most of the limestone river bluffs occur along the Mississippi River or on the eastern side of the state. A few woody species are able to send their roots into the cracks and crevices of the bluffs. These plants, because of the extreme environmental conditions to which they are subjected, are usually gnarled and stunted. Included in this group are buckthorn (*Rhamnus caroliniana*) (Fig. 41), dwarf hackberry (*Celtis pumila*), red cedar, and blue ash (*Fraxinus quadrangulata*). Several vines are found clinging tenaciously to the exposed faces of many of the limestone river bluffs. Some of these are Virginia creeper, bur cucumber (*Sicyos angulatus*), snailseed (*Cocculus carolinus*), and two rare species, dutchman's pipevine (*Aristolochia tomentosa*) and *Calycocarpum lyonii*.

Only a relatively few herbaceous species occur under such rigorous conditions. The most common ones are cleft phlox and *Allium stellatum*, while plants associated with the talus include species of *Polymnia* of the Compositae.

Numerous calcicolous ferns occur, the most characteristic being walking-fern (*Camptosorus rhizophyllus*), black spleenwort (*Asplenium resiliens*), tiny lip-fern (*Cheilanthes feei*), purple cliff-brake (*Pellaea atropurpurea*), and bulblet bladder-fern (*Cystopteris bulbifera*). Other species, some of which are exceedingly uncommon, are woolly milkweed (*Asclepias lanuginosa*), *Isanthus brachiatus*, shrubby bluet (*Houstonia nigricans*), and species of blazing-star (*Liatris* spp.).

Lowland Grass Communities

Prairie is a French word having the meaning of meadow. Thus prairie is a plant community dominated by the grass life form. Many other herbaceous non-grass plants are also present, living an adjusted life of subordination to the grasses. Treelessness is the rule, but under certain conditions some shrubs and trees are to be found. Along rivers or deeper ravines where forest plants may find favorable conditions, they extend themselves westward into the grasslands.

From Manitoba, Canada, to Texas and east of the 99th meridian lies the eastern triangular wedge of the grassland area known as true prairie. The true prairie descends the western and southern edge of Minnesota, enters Illinois in the driftless area of the northwest, and crosses the northern part to form a rounded lobe at the eastern edge of Illinois. Outliers of the true prairie exist farther eastward in Indiana and Ohio, being called a prairie peninsula by Transeau (1935). From its eastern edge the prairie extends from south of Lawrenceville southwestard to a point south of St. Louis, Missouri. Finger-like extensions of prairie remnants run into the forested uplands and better drained lowlands thus forming a mantle of grassland with a tattered southern edge. South of this zone prairie areas occur less frequently, usually becoming more isolated and increasingly smaller southward. Only a few natural prairie situations are now known in southern Illinois, a region which originally was largely forested. Former prairie locations within southern Illinois are indicated by such place names as Burnt Prairie in White County, Belle Prairie in Hamilton County, Prairie, and Prairie du Rocher in Randolph County, Herrin and Davis Prairie in Williamson County, and Elk Prairie in Jackson County.

To early explorers and settlers having a woodland heritage the prairie was a strange and uncertain entity. The immensity of unobstructed land and sky, the waves of grasses rippling before the winds as on a sea, and the brilliant adornment of seasonal herbs are all mentioned time and again in early accounts with poetic emotion. A heavy impenetrable sod and strong masses of roots, the secret of unmatched but then unsuspected fertility, was a great reason for delay in the settlement of the prairies. There was a fear of becoming lost in its vastness, and fear of the fires which periodically occurred.

It is quite understandable in the light of these impressions that this strange treelessness was an immediate object for explanation. Most early explanations were made in terms of fire. This explanation came easily for those who had witnessed the spectacle of a prairie fire. Such a fire was an awesome, fearful, though from a safe vantage point, an indescribably beautiful thing. Above all else it formed a lasting impression, and certainly everyone understood it as an enemy of wood or other combustible material. Fire often occurred from lightning strikes and was known to have been purposefully set by Indians to improve hunting or by the white man to improve visibility or remove an impediment to his travel.

Later explanation of treelessness focused on the thinness or fine texture of the soil, or the influences of bison and other grazing animals through their trampling and grazing pressures. Some held that a lack of mycorhiza (a fungus-root relationship) was the answer. Climate as the obvious cause has recently been given more sophisticated inquiry. Having a geographic midcontinental location the prairie receives the extremes of heat and dryness in summer and cold in winter. Summer air masses passing over the prairie region in summer bring less rain, lowered humidity, higher temperatures, higher rates of evaporation, and a higher incidence of drought than is experienced in forested areas farther eastward; hence, the area is more favorable to the grass life form.

The general climatic conditions earlier enumerated correlate with the main body of prairie, but do not fully explain the distribution of prairie along the humid edge of the grasslands or still farther eastward where the existing climate supports both trees and grassland. In these transition regions it will not be argued that fire doesn't help to suppress trees, but in locations where such factors as soil depth drainage and slope aspect produce a topographically dry situation the trees are in a tenuous balance and a disturbance of most any kind causing an opening of the soil would or might permit the entry of prairie plants. Such invasions are known in the edge of the forest where clearing for roads or telephone lines has been made. A change in climate sufficient to cause the death of trees on exposed thin soil areas accomplishes the same result, i.e. a possible replacement of trees by grasses. The question to be answered is at what time the replacement occurred and what is the present stability of the grassland. Some present hill prairie areas give the appearance of having been there for a long period of time—perhaps thousands of years. Evers (1955) has stated "The growth of grassland rather than forest on the upper bluff slopes is attributed to priority of occupation by prairie species and to the xeric conditions that are produced by the combination of local exposure to the sun and to the wind (especially to wind moving unimpeded across wide floodplains), the height of the bluffs above the adjacent bottomlands, the steepness and direction of the upper slopes, and the permeability of the substratum. Thus, the hill prairie community is the result of a complex set of conditions, the effectiveness of which is determined by location and topography.

The success of grasses in such an environment may be attributed to their remarkable adaptation structurally and physiologically. They differ from other plants in their growth form by having at the base of the leaf a meristematic or growth zone which allows for the regrowth of the blade tissue even if bitten off above. New growth from the meristematic zone elongates the blade in much the same fashion as lead is raised in a mechanical pencil, i.e., the new part comes from below. As grasses may have their stems as well as their leaves bitten off, growth may also come from buds in the axils of sheathing leaves at the base of the stem. Unless grazing is extremely close these buds are not destroyed. Buds which burst through the basal leaf or sheath and develop in a lateral direction are extravaginal (out of the sheath) buds. Where these buds develop below ground they give rise to a rhizome or underground stem and are well-protected from fire, grazing, or trampling. If lateral growth occurs above ground, a stolon results. Such a species as buffalo grass (*Buchloe dactyloides*) of the western prairies is an example of the stoloniferous habit. Buds remaining in the sheath are intravaginal and all growth is more or less vertical, thus resulting in bunch grasses. Such species as needlegrass (*Stipa spartea*) and June grass (*Koeleria cristata*) are examples and these are apt to have damage to their buds from grazing.

Grasses have an ability to grow at their nodes. Stems which are bent and broken down or lodged can straighten or right themselves by a differential rate of growth from the shaded side of the stem node. This growth is due to a hormonal stimulus.

Grass leaves are vertical and well displayed to available sunlight. They are flexible and the whole plant bends, therefore not suffering from winds. The internal anatomy of the grass leaf blade is a wonder to behold. Bridge-like spans of epidermis and chlorenchyma are supported by "I" beams and "T" beams that would thrill an engineer or architect. Some of our most important discoveries we only find to be copied from nature. Edison once said "until man makes anything as wonderful as a blade of grass, nature can laugh at his efforts." Some species with an abundance or preponderance of sclerenchyma are termed hard grasses, others with an emphasis upon chlorenchyma are known as soft grasses. The amounts of soft or hard tissues and water content and structural design reflect the fitness of the different species for the environments in which they are normally found.

Leaves of grass plants also form a great amount of litter for the development of a protective mulch. This layer of dead and decaying leaves insulates the soil keeping down the summertime surface temperatures and lowering surface evaporation. Impact of raindrops is lessened and the infiltration of rain is promoted so that runoff and erosion are reduced.

A final way in which grasses are successful is that their roots firmly hold the soil against erosion. Many species have deeply penetrating roots which are intimately in contact with the soil (Weaver, 1934).

If one could still have the experience of walking the undulations of undisturbed true prairie landscape, there would be revealed over and over again a natural assortment of dominant species or communities of grasses in a more or less definite ordination according to moisture gradients. Upon entering the lowlands, particularly in the early part of the year, there would be water of perhaps shoetop depth. Slough grass prevails in this situation in nearly pure stands. Ascending the slopes onto overflow bottoms, one encounters switch grass. On better drained but moist areas is Canada wild rye. Then on the better drained moist lower slopes are big bluestem and Indian grass. These communities are characterized by their tallness and their sod-forming habit.

The upland species are likely to be little bluestem, side-oats grama, prairie dropseed, needlegrass, and June grass. These species are middle-sized grasses. They are bunch-forming or scattered and tufted between other species in an interstitial manner. This ordination of communities is in places strikingly vivid and in other places more obscure because of the lack of definition of particular habitats and because of the greater tolerances or less exacting demands of certain species (Weaver, 1934). A comprehensive and very understandable treatment of the true prairie is given by Weaver (1954) in his book *North American Prairie.*

The former finger-like extensions of prairie into southern Illinois today are reduced to long ribbons of prairie vegetation found only along highways and railroad rights-of-way. In these places, the disturbing influences, chiefly burning and mowing, are a part of the management program in maintaining the right-of-way. Unfortunately, overzealous weed control efforts have reduced the prairie vegetation, the natural mulch has been destroyed, and nesting sites and food of animals have been eliminated. Thus in many places, there occur areas which are, comparatively speaking, biological wastelands. The loss of important species of plants and animals is slow to be recognized, and the loss is then belatedly lamented.

Big bluestem, at the time of settlement, characterized the mesic prairies perhaps more than any other species. Today the railroad and highway right-of-way strips are dominated by Indian grass which usually indicates a mild disturbance such as frequent burning. Prairie vegetation along highway 13 north of Murphysboro, intermittently as far as Coulterville, shows a predominance of Indian grass. This is also true for the prairie vegetation north of Carbondale on route 51 as far as DuQuoin.

Quadrat studies of relict prairie vegetation near Finney in Jackson County along highway 13 shows the presence of several weedy species such as wild lettuce (*Lactuca scariola*), *Eragrostis pectinacea*, yellow foxtail (*Setaria lutescens*), and purple-top (*Triodia flava*). A lower basal area of 7.6 per cent rather than the usual 13–15 per cent for an Indian grass or bluestem community is indicative of disturbance (Table 27). Indian grass at this location, along with big bluestem, had excellent height growth as many specimens were measured between 237 and 296 centimeters.

Herbs are conspicuous during all seasonal aspects and in the main keep pace with the grasses in their height growth. The fall blooming herbs of such species as the rosin-weeds (*Silphium* spp.— Fig. 42), blazing stars (*Liatris* spp.), and rattlesnake-master (*Eryngium yuccifolium*) are nearly as tall as the grasses and are quite conspicuous.

During the vernal aspect the outstanding species include puccoon (*Lithospermum canescens*), blue-eyed grass (*Sisyrinchium* spp.), spiderwort (*Tradescantia ohiensis*), and wild hyacinth (*Camassia scilloides*). A low shrubby willow (*Salix humilis*) is also frequently

27. *Percentage composition of vegetation in 8 one-meter quadrats from a railroad right-of-way prairie strip*

SPECIES	1	2	3	4	5	6	7	8	Average
Sorghastrum nutans	89.76	....	10.10	3.18	....	74.60	17.31	91.22	35.77
Andropogon scoparius	4.42	98.40	78.80	71.79	88.40	....	....	....	42.72
Andropogon gerardii	0.02	....	0.77	....	....	....	....	....	0.08
Lespedeza virginica	3.26	....	3.20	....	....	....	5.32	3.02	1.75
Solidago altissima	0.72	....	....	....	....	....	....	....	0.09
Rubus frondosus	1.73	0.40	....	0.06	....	....	....	....	0.17
Lactuca scariola	0.07	....	....	....	....	....	....	....	0.00
Eragrostis pectinacea	....	1.20	0.30	....	9.35	....	....	....	1.25
Carex spp.	....	....	5.48	18.59	....	....	2.01	5.35	3.92
Erigeron strigosus	....	....	0.36	....	....	....	....	....	0.04
Tridens flavus	....	....	1.10	....	....	....	....	....	0.13
Panicum spp.	....	....	....	0.65	1.02	....	....	....	0.20
Setaria lutescens	....	....	....	3.29	....	....	....	....	0.41
Apocynum sibiricum	....	....	....	2.70	....	....	....	....	0.34
Acalypha lanceolata	....	....	....	....	1.12	....	....	....	0.14
Oxalis stricta	....	....	....	....	0.01	....	....	....	0.00
Sporobolus asper	....	....	....	....	....	....	75.36	....	9.42
Poa pratensis	....	....	....	....	....	23.80	....	....	2.97
Elymus virginicus	....	....	....	....	....	0.72	....	....	0.09
Muhlenbergia mexicana	....	....	....	....	....	0.68	....	....	0.09
Viola spp.	....	....	....	....	....	....	....	0.42	0.05
Desmodium spp.	....	....	....	....	....	....	0.87	....	0.11
BASAL AREA	13.00	11.31	11.04	9.37	3.20	2.56	5.02	5.79	7.66

Data collected near Finney, Jackson County, 1961.

seen. Common summer herbs include culver root (*Veronicastrum virginicum*), fever few (*Parthenium integrifolium*), and others.

Other biota of the mesic prairie community of the rights-of-way include meadowlarks, quail, rabbit, prairie king snakes, grasshoppers, and spiders. In less disturbed areas, prairie mice are usually found.

Slough Grass

Spartina pectinata is a coarse grass reaching a height of six to eight feet. Its foliage obscures the ground, but the area occupied by its stems near the ground (basal area) is only one to three per cent. Slough grass forms a sod, and because its growth and flowering is so much in the warmer part of the year, it is known as a warm season grass and its ancestry assumed to be southern. Slough grass is well adapted to a soil condition of poor aeration and rarely has to share its habitat with other grasses. Thus it forms nearly pure stands. These soils are too wet for cultivation until drained.

A slough grass community may often be called wet prairie and, because the single dominant is a grass, this is somewhat justifiable. The slough grass community is, in succession, however, a last stage

42. Four species of rosin-weed are found in southern Illinois as prairie remnants: [left] prairie dock (*Silphium terebinthinaceum*); [right] cup plant (*S. perfoliatum*).

THE UPLAND SERIES

in the hydrosere. Preceding it as a wetter stage is the sedge-meadow, dominated not by grasses but by sedges, rushes, and other marshland plants. This condition is not to be considered as grassland, but as a stage leading to it. Wet prairie, infrequently seen in southern Illinois, was once very abundant in northern and central Illinois before the introduction of tile drainage in fields. Wet prairie has been found to be similar in nature to that of neighboring states, a not too surprising fact owing to the uniformity of the wet habitat.

Switch Grass

Panicum virgatum may be found in very moist lowland situations with but slightly better drainage than that of slough grass with which it often alternates. The basal area is again quite low, attaining only about five per cent, though its foliage is rank enough to completely obscure a view of the ground. Under these conditions there is little understory near the ground later in the season. Understory plants complete their growth early before being overtopped by the grasses. Switch grass is also a sod former and a warm season grass. It attains heights of three to six feet.

Fig. 42, contd. [left] compass plant (*S. laciniatum*); [right] rosin-weed (*S. integrifolium*).

Wild Rye

Elymus canadensis makes similar demands upon the habitat to those of switch grass with which it sometimes intermingles. In other instances, it forms small pure stands locally. Its presence is usually indicative of some minor disturbance. Wild rye is a cool season grass, forming a loose sod. Because its rhizomes are loosely arranged or in a thinner network, it is not too successful with other lowland tall grasses. Because of its early flowering and vegetative growth, it is assumed to be of northern extraction. It attains heights of four to five feet and has a basal area of five to eight per cent (Weaver, 1934, 1954).

Big Bluestem

Andropogon gerardii characterizes the true prairie more than any other species because of its widespread dominance. It is a tall grass attaining heights of six to eight or even ten feet and having twelve to fourteen nodes. The terminal inflorescence is two-, three-, or four-forked, causing it to sometimes be called turkey foot. Big bluestem occupies the moist bottomlands which have better drainage than any of the preceding types and in the eastern true prairie of Illinois, it dominated the uplands as well. Bluestem prairies were earlier described as being tall enough to hide a rider on horseback from view.

Vegetative growth is luxuriant and one-and-one-half to two feet high by mid-July. In mid- or late August it begins to flower and continues on into September. Thus, it is a warm season grass having migrated from the South. It is a sod-forming grass and an exceptionally good competitor. The basal area is usually about fifteen per cent. The causes of its dominance are attributed to its rapid development, dense rhizomes, great stature, and tolerance of its seedlings to shade (Weaver, 1934, 1954).

Indian Grass

Sorghastrum nutans is the last of five important lowland grasses. Interminglings of this grass occur with several other species. Indian grass does, however, make the same demands upon the habitat as does big bluestem. Both are warm season species in their growth and flowering and they are of comparable height.

Upland Grass Communities or Hill Prairies

Regions known as hill prairies occur near edges of bluffs in southern Illinois. The greater number of these are found on limestone rocks which border the Mississippi River. The appearance and composition of vegetation along these bluffs exhibit well-marked alternes. Hill prairies in southern Illinois have been recorded as early as 1868 when Worthen discussed loess-capped bluffs with grass-covered knobs. The appellation of hill prairie is attributed to Vestal by Evers (1955). The vegetation of these upland prairies is greatly different from that which is to be found on low ground or from that which may be found on level upland of the till plains of the most recent glaciation.

In the hill prairies, a bunch grass habit prevails. This contrasts with a predominantly sod-forming habit on level lowlands. The dominant grasses of the hill prairie are those of middle-sized stature which are most common on uplands and consist of little bluestem (*Andropogon scoparius*), June grass (*Koeleria cristata*), and side-oats grama (*Bouteloua curtipendula*). In most hill prairies a small amount of the tall grasses, big bluestem (*Andropogon gerardii*), and Indian grass (*Sorghastrum nutans*) is to be found. (For examples see Fig. 43.)

These dry prairies are nearly always bordered by shrubs such as sumac, rough-leaved dogwood (*Cornus drummondii*), and prairie crab apple (*Malus ioensis*). During certain wet years the invasion of seedlings of these species seems to progress, but when examined later when normal climatic conditions or a drought occurs, the seedlings are found to have died. There is, from year to year, some shifting in the populations but there is return to a former balance in a display of dynamic stability.

Species composition of the hill prairies studied by Evers (1955) was found to be characterized by a high proportion of plants from the southeast of the United States, although nine were characteristic of western prairies and three of the Ozark plateau. Evers postulates that hill prairie stands in Illinois have possibly existed on these Mississippi River bluffs from Wisconsin or pre-Wisconsin glaciation to the present, and that the prairies will exist in these sites until a change in climate occurs which will provide for more moist conditions or until erosion has gentled the slopes. Evers' study was concerned with 61 hill prairies in Illinois, of which 10 occur in the southern four tiers of counties. In these 10, Evers reported 141 different species of vascular plants. A total hill prairie flora of 390 vascular plants was revealed in the study of the 61 hill prairies. In the true prairie a total flora of about 265 species prevails while Steiger (1930) found 237 species on a square mile of prairie near Lincoln, Nebraska.

Little Bluestem

Andropogon scoparius is a bunch-forming warm season grass from three to four feet in height. Its growth begins in mid-April, with

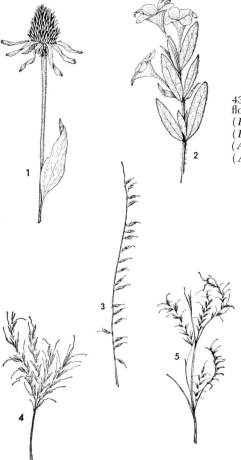

43. Plants of hill prairies: (1) purple coneflower (*Echinacea purpurea*), (2) wild petunia (*Ruellia humilis*), (3) side-oats grama (*Bouteloua curtipendula*) (4) big bluestem (*Andropogon gerardii*) (5) little bluestem (*Andropogon scoparius*).

THE UPLAND SERIES 157

flowering beginning in July, but not becoming complete until September. The bunches of little bluestem are six to eight inches in diameter and have several dozen flower stalks at flowering time except in dry years when it flowers only sparingly (Weaver, 1932). The bunches may otherwise have one or two hundred leafy stems. The distance between bunches is often a foot and the characteristic basal area in southern Illinois hill prairies has been found to be eleven per cent. From seed gathered in autumn and tested during successive months it was found that germination was highest in April, when it still was only five per cent (Blake, 1935).

Little bluestem was by far the most important species of any of the hill prairies examined. This importance is generally held on upland prairies and this species frequently makes up from fifty to ninety per cent of the vegetation. Little bluestem is easily recognized at a distance by the reddish-brown color it has during autumn. During its vegetative growth it may be recognized by its flattened stems and a whitish transverse line where the blade joins the stem.

Side-oats Grama

Bouteloua curtipendula is a species occurring in small bunches or tufts. It occurs sparingly intermixed with other vegetation to the extent that it is often called an interstitial species. This is a drought tolerant species and it is found in many southern Illinois hill prairies. It is most frequently found along and just back of the edge of a bluff in a very exposed situation or along a brow or shelf-like break in the west-facing hill-side. In such a situation side-oats grama may comprise most of the vegetation (Table 28, quadrat number 9).

Side-oats grama is common throughout true and mixed Prairie as well as in the Desert Plains Grassland and Texas Coastal Prairie Associations. The range of side-oats grama includes some of the more eastern states where this species is sparingly represented in rocky and dry places. Side-oats is a warm season grass of moderate size at flowering time when, with its flower stalks, it measures up to 2.5 or 3.5 feet in height. Leaves are from 3 to 10 or 12 inches long, drying to a curly and light tan condition in early fall. Flowering begins in July. The oat-like spikelets, 20 to 40 or more, arranged as they are on one side of the rachis, has caused this plant to be called side-oats grama. Best germination from fall-gathered seeds was obtained in February March, and April when it was 10 per cent (Blake, 1935).

28. *Percentage composition of vegetation in 12 one-meter quadrats from a hill prairie near Prairie du Rocher, Randolph County*

SPECIES	1	2	3	4
GRASSES				
Andropogon scoparius	81.20	72.20	64.80	86.70
Sorghastrum nutans	6.80	2.40	....	....
Bouteloua curtipendula	....	....	....	1.63
Muhlenbergia cuspidata	....	....	....	....
Sporobolus asper	....	....	....	....
Andropogon gerardii	....	....	....	....
FORBS				
Gerardia aspera	7.65	10.44	14.50	2.53
Aster oblongifolius	....	1.20	5.30	0.30
Lespedeza virginica	....	....	....	....
Echinacea pallida	1.34	11.20	1.76	....
Lespedeza capitata	0.10	....	....	7.09
Petalostemon purpureum	....	....	....	0.48
Solidago nemoralis	....	....	....	....
Linum sulcatum	0.72	....	14.50	1.27
Euphorbia corollata	0.12	....	....	....
Rudbeckia hirta	0.47	0.50	....	....
Sisyrinchium spp.	0.66	....	....	....
Liatris cylindracea	1.04	....	....	....
Ruellia humilis	....	....	....	....
Kuhnia eupatorioides	....	....	....	....
Physostegia virginiana	....	0.14	....	....
Cassia fasciculata	....	....	....	....
Rhus copalina	....	....	....	....
Croton monanthogynus	....	....	....	....
Houstonia nigricans	....	....	....	....
Carex spp.	....	....	....	....
Senecio plattensis	....	....	....	....
BASAL AREA	7.59	5.27	3.46	13.40

Data collected four miles north of Prairie du Rocher, Ill., October 21, 1961.

Plains Muhly

Muhlenbergia cuspidata is a warm season bunch grass. It is a species of the west and central plains region. On the loess soils of the central Nebraska region it has been described as a species of secondary importance (Weaver and Albertson, 1956). This species also grows on the loess covered west-facing hills of southwestern Illinois. It was reported by Evers (1955) from a hill prairie at Bluff Springs in Cass

5	6	7	8	9	10	11	12	Average
63.30	61.50	81.40	44.40	2.99	84.90	77.06	14.60	61.25
1.62	27.50	....	....	....	....	....	62.73	8.42
....	....	1.43	34.28	69.90	....	....	....	8.93
....	....	....	....	12.32	....	....	....	1.03
....	....	....	....	10.74	....	....	....	0.89
....	....	....	....	1.41	....	....	....	0.12
1.70	2.25	7.14	....	....	3.88	5.41	....	4.62
7.20	2.63	2.46	1.75	1.20	0.24	1.64	9.78	2.80
13.16	2.29	2.30	2.55	....	8.22	15.65	6.27	4.20
....	....	1.38	....	....	0.47	....	4.00	1.67
1.08	0.67	....	....	....	0.47	....	....	0.78
8.50	3.17	2.15	5.32	....	....	....	....	1.63
1.13	0.34	0.51	2.30	....	0.59	....	....	0.40
1.19	....	1.38	....	....	....	....	....	1.60
0.54	....	....	0.23	....	....	....	....	0.08
....	....	....	....	....	....	....	....	0.09
....	....	....	....	....	0.59	....	....	0.10
....	....	....	....	....	....	0.12	....	0.10
....	....	....	....	0.12	....	....	0.13	0.02
....	....	....	0.06	....	....	....	2.40	0.21
....	....	....	....	....	....	....	....	0.01
....	0.07	....	....	....	....	....	....	0.01
....	0.13	....	....	....	....	....	....	0.01
....	....	....	....	0.12	....	....	....	0.01
....	....	....	....	1.20	....	....	....	0.10
....	....	0.41	....	....	....	....	....	0.03
....	....	....	....	....	....	0.12	....	0.01
11.30	17.90	14.84	7.20	6.02	15.00	14.50	12.60	11.17

County. In the hill prairie north of Prairie du Rocher, the plains muhly was accompanied by side-oats grama and *Mentzelia oligosperma* on the thin loess soil near the rim of a 300 foot limestone bluff in a very dry and exposed situation. Stems and foliage at maturity are from one to two-and-one-half feet high and are very slender, tough, and wiry. Its spikelets are single-flowered and the florets are shed as easily as are those of the dropseeds.

June Grass

Koeleria cristata is a cool season small-sized bunch grass. The bunches are from one or two inches in diameter up to four or five. At flowering time in June it stands two-and-one-half feet high. The spikelets crowd the inflorescence and look, in a way, like the heads of timothy. The amount of cover formed by June grass is low, perhaps less than five per cent of the composition, and it exists in several of our dry upland hill prairies. In the hill prairie the bunches have from only a few flower stalks to twenty or more. Leaves of this species become green in early spring. The leaves twist into a spiral growth as they become longer and they are unevenly ribbed on their dorsal surface.

Tall Dropseed

Sporobolus asper is a drought resistant species having a height of one-and-one-half to two-and-one-half feet. The stems are wiry and the leaves are short and possessed of much fibrous tissue. The inflorescence is two-thirds enclosed by a sheath even at maturity. In late fall and winter the short, strong leaves bleach and become frayed in the wind. Tall dropseed derives its name from the fact that its seeds are readily shed (dropped) as the inflorescence emerges.

Seasonal Aspects

Little flowering activity is apparent during the early part of the growing season. The hill prairie, a nondescript mass of reddish-brown color during February and early March, soon begins to awaken. Largely obscured by the ragged remnants of last season's growth, but visibly green on the soil surface are the less than one-inch wide rosettes of vernal whitlow-grass (*Draba verna*) and its taller kin, the wedge-leaved whitlow-grass (*Draba cuneifolia*), both in bloom during mid- or late March. During April, another rosette-forming species of diminutive proportion is *Androsace occidentalis* of the primrose family. These early-blooming species are annuals of small size, having also in common a copious production of flowers. Species of blue-eyed grass (*Sisyrinchium* spp.) also bloom in late spring. As the season progresses, the number of flowering species increases through early summer and then declines. There is a smallness of stature in the early blooming species. Later a correlation of herb height and grass height is observed so that the herbs bloom at or above the level of the

44. [left] Blazing star (*Liatris squarrosa*), a fall bloomer of upland prairie. [right] Purple prairie clover (*Petalostemum purpureum*), on the hill prairie at Government Rock in the Pine Hills.

grasses. In autumn the herbs are coarse and tall and most of them bloom at a level equal to or overtopping the grasses.

In early summer the prairie presents a more vivid display of bloom. Showing bright pinkish-purple blossoms among its finely divided foliage is the wild vervain (*Verbena canadensis*). Lavender flowers of *Gerardia* and darker purple or blue or even rose color of spiderwort (*Tradescantia virginiana*) enliven the scene. Two milkweeds, the orange-flowered or butterfly weed (*Asclepias tuberosa*) and the green milkweed (*Acerates viridiflora*), blend their rich hues to the floral display. Spatterings of white from the flat-topped spurge occur commonly throughout the hill prairies. An occasional white-blooming herb of early summer is *Anemone virginiana*. As the summer progresses, members of the legume family become noticeably conspicuous. Among them are both the purple and white prairie clovers (Fig. 44), partridge-pea (*Cassia fasciculata*), milk-pea (*Galactea volubilis*), species of lespedeza (*Lespedeza virginica* and *L. capitata* among others), and beggar's-ticks (*Desmodium* spp.). Coneflowers (*Echinacea pallida*) also contribute to the summer coloring among the grasses.

Fig. 44, contd. A copious mulch from a meter quadrat in hill prairie at Government Rock, Union County.

The autumnal aspect is marked by two features. It is the time of composites (asters, rosin-weeds, and their kin) and the predominating color is yellow. It is as if black-eyed susans (*Rudbeckia hirta*), goldenrods (*Solidago* spp.) and rosin-weeds are the reflection of a lingering and fading summer sun. By late October only darkened and dried heads of lespedeza and black-eyed susans remain along with the silvery pappus of the boneset (*Kuhnia eupatorioides*) and blazing star (*Liatris* spp.).

Prairie du Rocher

Both north and south of the village of Prairie du Rocher, a distance of five to six miles or more, exists a massive limestone bluff which varies in relief up to approximately 300 feet. The bedrock is covered with light yellow-brown, uniform-textured, and fine-particled loess soil. The loess varies in thickness, but for the most part it is thin on these undulating west-facing bluffs. Because of the undulation in the topography, both northwest and southwest slope aspects are presented. In the valleys or coves between northwest-facing and southwest-facing prairies exist such woody species as dogwood, ash, and sumac.

The hill prairie slopes from its highest point at the wood's-edge to the edge of the bluff over a distance of 50 or 100 yards. The slope varies from 10 to 30 or even 40 degrees. Thus upon west-facing or southwest-facing slopes, a very dry climatic and soil condition results.

Little bluestem is the leading grass species, making up 61 per cent of the composition (Table 28). The little bluestem here was sparingly in flower, though the bunches were numerous and otherwise well-developed. The stature of the plants was measured by about two-and-one-half feet of growth. The reduction in number of flower stalks was occasioned by the very dry latter half of the 1961 growing season. Basal area in the little bluestem community of this hill prairie averaged eleven per cent. Scattered amounts of Indian grass and to a lesser extent some big bluestem was found in the ravines and on the north-facing side of the prairie. Side-oats grama was scattered and sparingly represented. Its distribution is usually as an interstitial species except in places where a break in the topography exists such as a brow of a hill or a shallow step or shelf. In such places side-oats grama reaches a better development and a higher percentage of composition. Along the margins or the bluff edge and back to a distance of five or six feet, side-oats grama formed a distinct community (Table 28, quadrat 9). Here side-oats made up nearly seventy per cent and its companions were plains muhly and tall dropseed (*Sporobolus asper*).

Old Stone Face

This weathered countenance, a product of nature's artful sculpture, has witnessed many events in the passing of ages. Old Stone Face rests in the solitude of its scrubby trees of post oak, black jack oak, red cedar, and Buckley's hickory. A small hill prairie has kept it company for thousands of years. Here a narrow strip of prairie about 12 to 20 feet wide parallels the edge of the bluff for several yards (Fig. 45). The prairie exists on the thin deposit of loess occurring over sandstone bedrock. The exposure is nearly directly west.

45. A small hill prairie atop the bluff near Old Stone Face in Saline County has doubtless existed since before continental glaciation approached this area.

29. *Percentage composition of vegetation in 10 one-meter quadrats from a hill prairie in Saline County*

SPECIES	1	2	3	4
Koeleria cristata	1.60	22.60	....	....
Andropogon scoparius	36.28	68.00	90.00	82.35
Antennaria plantaginifolia	....	....	....	6.43
Verbesina alternifolia	32.10	....	....	....
Sorghastrum nutans	22.74	4.04	2.14	....
Petalostemum candidum	....	....	4.59	7.14
Helianthus divaricatus	....	....	....	....
Tephrosia virginiana	....	....	....	....
Andropogon gerardii	....	....	....	....
Danthonia spicata	....	....	....	....
Lespedeza virginica	....	4.20	....	....
Liatris squarrosa	....	....	....	....
Plantago spp.	....	....	....	....
Solidago spp.	....	0.30	1.22	....
Rosa setigera	....	....	....	....
Muhlenbergia capillaris	....	....	....	....
Gerardia purpurea	0.30	....	....	....
Cassia fasciculata	....	....	2.01	2.04
Elymus villosus	3.47	....	....	....
Aristida dichotoma	....	....	....	....
Crotalaria sagittalis	....	....	....	....
Diodia teres	....	....	....	....
Heuchera spp.	....	....	....	....
Cunila origanoides	....	....	....	....
Cercis canadensis	1.73	....	....	....
Hypericum gentianoides	....	....	....	....
Panicum depauperatum	....	....	....	....
Acalypha lanceolata	....	....	....	....
Tridens flavus	1.04	....	....	....
Muhlenbergia sobolifera	....	....	....	....
Lechea tenuifolia	....	0.90	....	....
Fragaria virginiana	0.87	....	....	....
Sporobolus vaginiflorus	....	....	....	....
Gerardia skinneriana	....	....	....	....
Lespedeza stuvei	....	....	....	....
Dodecatheon meadia	....	....	....	....
Sporobolus asper	....	....	....	....
Aster spp.	....	....	....	2.04
Aster oblongifolius	....	....	....	....
BASAL AREAS	2.40	6.13	8.20	8.90

Data collected at Old Stone Face, 1961.

5	6	7	8	9	10	Average
22.64	22.96	....	5.46	70.43	....	14.56
20.41	54.98	77.93	46.20	11.98	84.25	57.23
3.84	....	....	18.25	15.39	3.00	4.69
....	....	....	....	....	....	3.21
....	....	....	....	....	....	2.89
....	10.00	3.41	....	....	....	2.51
....	....	11.45	....	....	6.57	1.80
....	....	0.85	....	....	....	0.08
....	....	....	21.63	....	....	2.16
20.44	....	....	....	....	....	2.04
8.51	....	....	....	....	....	1.27
0.89	....	....	8.16	0.13	....	0.92
....	7.37	....	....	....	....	0.74
5.15	....	....	....	....	....	0.67
3.19	....	2.92	....	....	....	0.61
....	....	....	....	....	6.18	0.62
5.02	0.38	....	....	....	....	0.57
....	....	....	....	0.13	....	0.42
....	....	....	....	....	....	0.35
3.30	....	....	....	....	....	0.33
2.70	....	....	....	....	....	0.27
....	1.40	0.85	....	....	....	0.23
....	1.79	....	....	0.05	....	0.18
....	....	1.74	....	....	....	0.17
....	....	....	....	....	....	0.17
1.60	....	....	....	....	....	0.16
1.60	....	....	....	....	....	0.16
....	....	....	....	1.33	....	0.13
....	....	....	....	....	....	0.10
....	....	0.85	....	....	....	0.09
....	....	....	....	....	....	0.09
....	....	....	....	....	....	0.09
0.71	....	....	....	....	....	0.07
....	0.38	....	....	....	....	0.04
....	0.43	....	....	....	....	0.04
....	0.29	....	....	....	....	0.03
....	....	....	....	0.20	....	0.02
....	....	....	....	....	....	0.20
....	0.02	....	....	....	....	0.00
7.05	5.54	2.80	5.66	4.63	2.66	4.82

Little bluestem is the leading dominant having a percentage composition of about 56 per cent (Table 29). June grass occurred frequently in small tufts. It comprised nearly 12 per cent of the composition. Wild rye and Indian grass were sparingly present. The prairie here had a rather poor structure as evidenced by the relatively low basal area of 4.8 per cent. Basal area in the other hill prairies each dominated by little bluestem averaged about 11 per cent. This is in contrast to a basal area of about 13 per cent in the little bluestem communities of the True Prairie of Nebraska. Seasonal production, averaging 81 grams per meter, was slightly less than half that of the prairie at Government Rock in Union County, Mulch averaged 450 grams per quadrat which was slightly heavier than that at Government Rock.

Herbs which occur in some abundance are blazing star (*Liatris squarrosa*), whorled milkweed *Asclepias verticillata*), white prairie clover, partridge pea, gerardia (*Gerardia skinneriana*), wild petunia (*Ruellia humilis*), mountain mint (*Pycnanthemum flexuosum*), and flat-topped spurge (*Euphorbia corollata*).

Government Rock

Government Rock hill prairie, located in the Pine Hills of Union County, may serve as another representative hill prairie. Here the prairie is in the form of a spur which extends from the summit of a slope down to rock cliffs, popularly called the "chalk cliffs" of southern

30. *Percentage composition of vegetation in 10 one-meter quadrats from a hill prairie in Union County*

SPECIES	1	2	3	4
Andropogon scoparius	99.40	87.65	73.12	81.62
Bouteloua curtipendula	....	6.23	....	13.98
Andropogon gerardii	0.60	1.09	1.12	....
Sorghastrum nutans	....	5.03	25.76	4.38
Aster oblongifolius	....	....	....	....
Petalostemum purpureum	....	....	....	....
Cornus drummondii	....	....	....	....
Echinacea pallida	....	....	....	....
Lespedeza capitata	....	....	....	....
BASAL AREAS	19.30	1.99	5.32	7.19

Data collected at Government Rock, Pine Hills, Union County, October 1961.

THE UPLAND SERIES

Illinois. At the base of the spur, which faces west to southwest, is an abundance of scattered chert stones. Between this cherty base and summit of the slope is the hill prairie.

A scarcity of taller grasses is observed at once and conversely there is a greater amount of little bluestem and side-oats grama. Intermixed with these two dominants are scattered open bunches of big bluestem and indian grass in the more favorable places. Little bluestem makes up 82 per cent of the composition and side-oats grama about 6 per cent (Table 30). Basal area of this prairie was determined to be about 11 per cent in the community dominated by little bluestem. The soil, typically prairie in nature, was granular, porous, with good crumb structure, black with organic matter, and about two feet deep over limestone bedrock. It was well covered with mulch. Seasonal production averaged 145.8 grams per square meter and mulch which was accumulated from many seasons averaged 353.8 grams per square meter (Fig. 37).

Annotated List of Hill Prairies

Belle Smith Springs	Sect. 33, T 11 S, R 1 E	Pope Co.
Cave Creek	Sect. 28, T 13 S, R 3 E	Johnson Co.
Fountain Bluff	Sect. 36, T 9 S, R 4 W	Jackson Co.
Fults	Sect. 9, T 5 S, R 9 W	Monroe Co.
Government Rock	Sect. 9, T 11 S, R 2 W	Union Co.
Old Stone Face	Sect. 16, T 10 S, R 7 E	Saline Co.
Pine Hills North	Sect. 33, T 10 S, R 3 W	Jackson Co.
Prairie du Rocher	Sect. 32, T 5 S, R 9 W	Randolph Co.

5	6	7	8	9	10	Average
78.14	91.93	84.69	80.78	68.85	76.15	82.23
....	....	....	5.73	24.81	13.12	6.38
....	4.32	....	1.32	....	2.13	1.05
....	3.60	15.31	4.55	5.90	5.70	6.24
21.86	....	....	0.25	0.44	....	2.35
....	....	....	3.60	....	2.90	0.65
....	....	....	2.37	....	....	0.24
....	....	....	1.40	....	....	0.14
....	0.15	....	....	....	....	0.02
6.42	6.72	19.70	14.50	6.25	4.58	10.98

Upland Forests

Whereas topographic variation described for the lowland river series is often obscure, the reverse is true of southern Illinois uplands. These diverse physical conditions result in a multitude of environmental contrasts and nature has responded with as many vegetational patterns. These have been slowly, carefully, and selectively produced in the variously afforded habitats. In all but the best defined habitat areas, the result is expressed by subtle combinations of plants. It is through these best defined topographically related habitats that we shall make our introduction to the forest communities of upland.

In ravine bottoms occur many species of trees (Table 31). Those which associate with each other most often and which occur in greatest numbers are beech, tulip tree, and maple. Thus a recognizable community of these dominants exists. On slopes of hills are found communities of oaks and hickories. These are variable in composition, but white oak, red oak, and black oak are pre-eminently the leading species. These species of oaks and hickories continue to the crests where dryness induced by greater exposure presents a community of post oak and black jack oak (Fig. 46). These communities from ravine to ridge occur and reoccur all over southern Illinois where suitable habitats are found. There are, along this topographic gradient, many interminglings or associations of tree species to form still other assemblies (Table 32).

Fagus-Acer saccharum-Liriodendron/Rhus radicans *Community*

From ravine bottoms and the lowest north-facing slopes occurs a community of beech, sugar maple,* and tulip tree (Table 33). Beech

---

* Considerable difficulty is encountered in the exact status of the sugar maple in extreme southern Illinois. In alignment with the study by Desmarais (1942), the taxa which occur are as follows:

Acer saccharum ssp. *saccharum*—common in woods; leaves of moderate size, nearly glabrous below.

Acer saccharum ssp. *schneckii*—our most common form; leaves of moderate size, densely hairy below. This is often mistaken for *A. nigrum*.

31. Percentage of frequency and composition and number of lowland areas in which selected species were found

SPECIES	Frequency	Composition	Areas
Fagus grandifolia	16.70	16.25	17
Acer saccharum var. saccharum	34.96	14.96	20
Quercus alba	21.05	12.51	17
Quercus rubra	14.33	10.55	19
Liriodendron tulipifera	14.70	9.66	20
Nyssa sylvatica	9.25	5.01	18
Quercus velutina	7.04	4.97	17
Carya glabra	9.66	3.97	13
Carya ovata	8.75	3.96	16
Fraxinus americana	5.67	3.37	14
Carya cordiformis	5.26	2.93	15
Ulmus rubra	5.46	2.84	11
Ulmus americana	2.32	1.77	8
Quercus muhlenbergii	2.92	1.71	12
Juglans nigra	3.04	1.38	12
Liquidambar styraciflua	0.83	1.00	3
Platanus occidentalis	1.11	0.63	14
Carya laciniosa	1.00	0.34	4
Sassafras albidum	0.96	0.33	5
Crataegus spp.	0.17	0.28	1
Celtis occidentalis	0.67	0.26	3
Tilia americana	0.65	0.24	2
Juglans cinerea	0.33	0.24	1
Populus deltoides	0.17	0.17	1
Morus rubra	0.44	0.16	3
Carya tomentosa	1.04	0.13	4
Ostrya virginiana	1.04	0.13	6
Carya ovalis	0.33	0.11	2
Acer rubrum	0.38	0.10	2
Quercus falcata	0.17	0.09	1
Quercus macrocarpa	0.17	0.06	1
Cornus spp.	0.44	0.05	2
Carpinus caroliniana	1.42	0.01	4
Diospyros virginiana	0.15	0.01	2
Juniperus virginiana	0.15	0.01	1
Prunus serotina	0.17	0.01	1

Data from 20 areas.

Acer saccharum ssp. *floridanum*—swampy woods known only from a single collection in Union County; leaves small, densely hairy below. No attempt is made to distinguish the subspecies of sugar maple in the general discussion.

32. *Percentages of dominants in lowland, midslope, and upland communities*

Location	Lowland
Kaskaskia	43 Tulip, 18 S. Maple
Jackson Hollow	25 W. Oak, 22 S. Maple
Little Grand Canyon	53 Beech, 9 S. Maple,
Lusk Creek	22 W. Oak, 17 R. Oak, 15 S. Maple
Dry Hill	21 S. Maple, 11 R. Oak
Pine Hills	26 Beech, 16 Tulip
Fountain Bluff	30 Beech, 11 Tulip
Ferne Clyffe	17 Beech, 15, S. Maple, 11 B. Oak
Giant City	25 S. Maple, 13 R. Elm
Pomona	50 Beech, 20 W. Oak
Dixon Springs	15 Beech, 13 B. Oak, 13 R. Oak
Devil's Kitchen	34 R. Oak, 18 Tulip
Hooven Hollow	15 W. Oak, 12, S. Maple, 11 R. Oak
Panther's Den	35 S. Maple, 11 Sweetgum, 9 R. Oak
Blind Hollow	15 Sweetgum, 15 P. Hickory, 14 B. Oak
Beartrack Hollow	21 S. Maple, 15 Tulip. 9 Beech
Belle Smith Springs	51 Beech, 26 S. Maple
Pounds Hollow	25 R. Oak, 20 S. Hickory, 10 Tulip
Grindstaff Hollow	29 Beech, 26 S. Maple, 11 Tulip
Midland Hills	38 W. Oak, 18 Tulip

Key to abbreviations: Beech, R[ed] Elm, P[ignut] Hickory, S[hagbark] Hickory, Juniper, S[ugar] Maple, B[lack] Oak, B[lack]j[ack] Oak, P[ost] Oak, R[ed] Oak, S[humard's] Oak, W[hite] Oak, Sweetgum, Tulip.

33. *Percentage composition of trees in ravine forest communities*

SPECIES	Grandstaff Hollow	Belle Smith Springs	Kaskaskia Exp. Forest
Fagus grandifolia	28.53	50.71	5.48
Liriodendron tulipifera	11.43	5.46	43.12
Acer saccharum	28.50	26.42	17.95
Carya ovata	4.52	2.27	1.54
Carya tomentosa	1.58	....	....
Carya glabra	....	3.38	4.87
Carya laciniosa	2.26	....	....
Carya cordiformis	3.32	....	1.30
Quercus velutina	8.73	1.71	0.91
Quercus alba	....	....	0.91
Quercus rubra	9.26	2.40	9.32
Nyssa sylvatica	3.91	7.51	3.29
Ostrya virginiana	0.70	....	....
Carpinus caroliniana	....	0.11	....
Fraxinus americana	....	....	7.89
Juglans nigra	....	....	3.58

Mid-slope	Upland
25 R. Oak, 23 W. Oak	41 B. Oak, 23 P. Oak
42 B. Oak, 22 W. Oak	47 P. Oak, 23 Juniper
29 Beech, 22 W. Oak	47 W. Oak, 19 R. Oak
21 S. Hickory, 19 R. Oak, 15 W. Oak	68 P. Oak, 21 Bj. Oak
56 W. Oak, 23 B. Oak	43 W. Oak, 27 B. Oak
24 P. Hickory, 21 R. Oak, 20 B. Oak	33 W. Oak, 22 B. Oak, 20 P. Oak
41 W. Oak, 19 R. Oak	37 W. Oak, 21 R. Oak, 19 B. Oak
29 W. Oak, 21 R. Oak	48 Juniper, 25 P. Oak
66 B. Oak, 20 W. Oak	68 P. Oak, 21 Bj. Oak
40 W. Oak, 16 B. Oak	45 P. Oak, 25 W. Oak
17 W. Oak, 17 S. Hickory, 13 S. Oak	25 P. Oak, 23 W. Oak, 13 Bj. Oak
26 W. Oak, 25 B. Oak, 21 P. Hickory	45 P. Oak, 22 P. Hickory
....	64 P. Oak, 16 Bj. Oak
22 R. Oak, 19 W. Oak, 11 Tulip	39 P. Oak, 27 W. Oak, 14 B. Oak
32 W. Oak, 21 P. Hickory, 13 R. Oak	35 P. Oak 33 Bj. Oak
36 W. Oak, 16 B. Oak, 13 R. Oak	24 W. Oak, 19 B. Oak, 17 P. Oak
38 B. Oak, 35 W. Oak, 13 R. Oak	32 W. Oak, 25 P. Oak, 12 Juniper
39 W. Oak, 26 B. Oak, 20 R. Oak	54 P. Oak, 19 P. Hickory
16 Beech, 16 S. Hickory, 12 R. Oak	37 P. Oak, 17 P. Hickory, 13 W. Oak
38 B. Oak, 35 W. Oak, 20 R. Oak	26 B. Oak, 23 W. Oak, 22 P. Oak

34. *Association of species, random pairs of trees in lowlands*

NO. KEY AND SPECIES	1	2	3	4	5	6	7	8	9	10	11	12	13	14
1. Acer rubrum	0	0	1	1	0	0	0	0	0	1	0	0	0	0
2. Fagus grandifolia	0	27	10	7	5	12	18	5	6	4	5	6	1	2
3. Quercus alba	11	10	39	9	8	9	7	4	3	5	7	4	0	0
4. Quercus rubra	1	7	9	9	4	7	19	2	5	5	7	3	3	0
5. Quercus velutina	0	5	8	4	5	2	6	0	3	2	2	1	0	0
6. Liriodendron tulipifera	0	12	9	7	2	16	15	1	4	3	4	3	1	3
7. Acer saccharum	0	18	7	19	6	15	24	9	11	12	14	12	6	8
8. Fraxinus americana	0	5	4	2	0	1	9	0	3	0	1	0	3	1
9. Carya ovata	0	6	3	5	3	4	11	3	1	3	2	3	0	3
10. Nyssa sylvatica	1	4	5	5	2	3	12	0	3	3	2	0	0	1
11. Carya glabra	0	5	7	7	2	4	14	1	2	2	5	1	2	0
12. Carya cordiformis	0	6	4	3	1	3	12	0	3	0	1	1	2	0
13. Ulmus rubra	0	1	0	3	0	1	6	3	0	0	2	2	5	2
14. Juglans nigra	0	2	0	0	0	3	8	1	3	1	0	0	2	0

Data from 20 areas. Species found less than three times with other species omitted.

is usually abundant and stands in contrast to other species because of the distinctive gray color of its bark. It may form nearly pure stands locally or share dominance with maple, tulip tree, or white oak.

Here under copious shade, moisture is relatively greater than in other upland communities. The size of dominants, richness in variety of herbaceous species, and amount of accumulated litter emphasize a high degree of mesophytism. The affinities with mixed mesophytic forest types eastward are several.

The mid-layer of this community is composed primarily of dogwood, redbud, pawpaw, blue beech, spicebush, and, locally, the dwarf red buckeye. In the pre-vernal and vernal aspects these shrubs are especially noticeable and pleasing to the eye by virtue of color contrast between their flowers and drab color of canopy species whose leaves are not yet unfolded. Each community possesses a characteristic structure. This is revealed in the grouping of dominants, the kinds and abundance of mid-layer, and the herbaceous understory plants. The changes in topographic gradients are reflected by understory plants.

The ground layer of the beech-maple-tulip tree community is dominated by poison ivy in low shrub form. Some ivy also is found climbing in the trees, but this is less noticeable here than in floodplain communities where ivy is more common. Virginia creeper is nearly as abundant. A multitude of colorful herbs covers the forest floor during the vernal aspect. Among them are such better known plants as dutchman's breeches, spring beauty, squirrel corn, jack-in-the-pulpit, trout lily, wake robin, bluebells, windflowers, buttercups, and bloodroot.

Indian pipe (*Monotropa uniflora*), beech drops (*Epifagus virginiana*), and a rich fern flora which includes Christmas fern (*Polysti-*

46. Plants of upland forests: (1) broomsedge (*Andropogon virginicus*), (2) smooth sumac (*Rhus glabra*), (3) post oak (*Quercus stellata*),

chum acrostichoides), maidenhair fern (*Adiantum pedatum*), broad beech fern (*Dryopteris hexagonoptera*), sensitive fern (*Onoclea sensibilis*), and narrow-leaved spleenwort (*Athyrium pycnocarpon*), emphasize the amount of moisture in this community. The grape fern (*Botrychium virginianum*) has been dubbed by lay naturalists as "ginseng finder" as both these plants may be found in deep shaded and organically rich soils of woods.

Dominant Species Association

In lowland areas it is observed that sugar maple entered into more associations than any other species (Table 34). Beech, tulip tree, and white oak also were associated with a number of other species, but to a greater extent the association was among these three species. Thus, the close relationship of the dominants of the forest into communities of trees is emphasized. It appears that in the mesic lowlands there is a tendency toward a community of sugar maple, beech, white oak, and tulip tree. All of these except white oak and tulip tree are very shade-tolerant species and are able to reproduce under a heavy canopy closure.

American Beech (Fagus grandifolia)

Beech is a majestic and romantically appealing tree having a compact crown with numerous fine branches, a straight trunk, and smooth gray bark. It matures in about 150 years. Old specimens may reach an age of 300 to 400 years. Heights of better than 80 feet are attained and diameters of 30 to 40 inches may be seen in southern Illinois. Here beech inhabits the moist tributary ravines and favors a north slope aspect. Shanks (1953) found that in Ohio, beech is favored by heavy, poorly aerated soils and developed almost pure

(4) persimmon (*Diospyros virginiana*), (5) sassafras (*Sassafras albidum*), (6) white oak (*Quercus alba*).

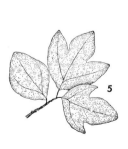

stands on such sites, while lighter, better drained, and aerated soils were favorable to sugar maple and resulted in beech-maple stands. It may be found only on south slopes where there is favorable surface moisture. Dense shade to which it is tolerant is an outstanding feature of its habitat. This tolerance makes it an inevitable member of late successional stages and climax on the lower ground.

Separate male and female flowers are borne on the same tree and appear with the leaves in late April or early May. Male flowers are borne in rounded heads on long stalks near the base of new growth. Female flowers are found in two- to four-flowered stalked clusters, and they, too, appear near the end of new growth. Fruit is formed during early autumn as a four-parted bur containing two three-angled nutlets. Large crops of seeds are produced at irregular intervals (Harlow and Harrar, 1958). Twigs are brown and bear slender pointed buds about an inch long. Leaves are alternately arranged, sharply pointed, coarsely toothed, thin but leathery, and up to five inches long (Fig. 4).

Tree species most commonly associated with beech include tulip tree, sugar maple, and sweetgum. To a lesser extent it may be found with elm, white oak, and sycamore.

Tulip Tree (Liriodendron tulipifera)

By reaching heights of nearly 200 feet, tulip tree becomes one of the tallest of eastern deciduous forest trees. It is a very straight and fast-growing tree which matures at about 200 years. Some specimens of a diameter rivaling sycamore have been known.

Dark green, four-lobed leaves which are four to six inches long appear from the oval, somewhat compressed, slightly two-edged buds in early April or mid-May. Flowers are two or three inches broad, having three greenish sepals and six light green petals each colored orange near its base. These perfect flowers appear in early May shortly after the appearance of leaves. Soon the perianth falls away to reveal a cone-like fruit two or three inches in length. These ripen in September or October, and the aggregate of samaras loosen to shed the individual terminally-winged seeds.

Seed crops vary from year to year and the average fertility is but fifteen per cent (Harlow and Harrar, 1958). Fluctuations in seed quality also vary from year to year. Seedling numbers resulting from equal quantities of seed vary as much among individual seed trees within a stand as from among stands of different geographic locations (Central States Forest Exp. Sta. Note 134, 1959). Seeds seem to germinate best in a mineral soil, hence are most abundant along logging trails and skid tracks where the forest floor has been scarified.

Tulip tree often forms recognizable communities in the deep moist bottoms of ravine forests or on terraces just above swamps and flood plains. Its most frequent associates are beech, maple, ash, basswood, elm, sycamore, and, to somewhat lesser extent, white, red, and black oaks.

Along slopes above ravine bottoms begins a gradual transition from the types below. Many of the same trees found on the lower half of the slopes are found in this area, but there is a change in the ranking of the dominants and the addition of still others. Although variable in composition, these oak forests have communities of trees expressed more subtly than those above or below them in the topographic gradient. Thus, to the casual observer, they have a homogeneous and uninteresting appearance.

Here, on an average from 20 areas of study, white oak made up nearly 30 per cent of the composition of the southern Illinois upland mid-slope forests. This species also had high frequency occurrence in all the samples taken with 47 per cent. Red oak held a percentage composition of 13 per cent and had a frequency occurrence of nearly 22 per cent. Black oak made up nearly 20 per cent of the composition while having a frequency occurrence of nearly 27 per cent. Taken together, the 20 sampling areas show a mid-slope forest composed of about 33 tree species. Among this number, sugar maple, white oak, and black oak were represented in all 20 areas and red oak was present in 19 of the 20 areas (Table 35). Fourteen species were present in only 1 to 3 of the areas and thus may represent chance distributions from different neighboring communities or possibly may be remnants from an earlier succession stage which was originally initiated by cutting.

Undergrowth is principally of seedlings of canopy species. In more advanced second growth, the following shrubs are usually present — dogwood, redbud, hop hornbeam, bladdernut, bittersweet, farkleberry, aromatic sumac, and New Jersey tea (*Ceanothus americanus*). Locally, as at Union County State Forest and the Pine Hills, on cherty west-facing slopes, may be found scattered bushes of *Rhododendron roseum* and *R. nudiflorum*.

35. *Percentage of frequency and composition and number of areas in which selected species were present on mid-slopes*

SPECIES	Frequency	Composition	Areas
Quercus alba	47.21	29.54	20
Quercus velutina	26.75	19.65	20
Quercus rubra	21.64	13.04	19
Carya glabra	15.91	7.97	18
Acer saccharum ssp. saccharum	20.98	5.24	20
Carya ovata	11.60	4.83	15
Fagus grandifolia	4.54	4.02	7
Liriodendron tulipifera	4.41	2.44	10
Nyssa sylvatica	4.50	2.42	13
Quercus muhlenbergia	1.48	1.30	6
Fraxinus americana	3.37	1.06	10
Quercus stellata	2.12	0.80	4
Quercus shumardii	0.17	0.65	1
Carya ovalis	0.67	0.51	2
Carya tomentosa	1.83	0.44	4
Liquidambar styraciflua	0.29	0.41	2
Ulmus americana	1.00	0.36	6
Juglans nigra	1.00	0.28	4
Carya buckleyi	0.17	0.24	1
Robinia pseudoacacia	0.17	0.21	1
Sassafras albidum	1.00	0.21	5
Ulmus rubra	0.67	0.21	4
Quercus marilandica	0.33	0.20	1
Ostrya virginiana	1.49	0.16	5
Carya laciniosa	0.46	0.11	2
Juniperus virginiana	0.17	0.05	1
Prunus serotina	0.17	0.05	1
Tilia americana	0.17	0.05	1
Acer rubrum	0.25	0.03	1
Cornus spp.	0.33	0.03	2
Ulmus alata	0.25	0.02	1

Data from 20 areas.

In examination of the herbaceous layer of wooded hillside communities, 600 meter quadrats were taken in 20 areas. Some species were found infrequently and were considered chance distributions or ephemerals and therefore of little importance. Only a few plants had a representation in as much as 70 per cent of all areas. These were three-seeded mercury (*Acalypha virginica*), tick-clover (*Desmodium* spp.), stone mint (*Cunila origanoides*), and the falcate violet (*Viola*

THE UPLAND SERIES

falcata). Other herbs which are conspicuous and generally characteristic of wooded slopes, but less faithfully represented in as many separate areas, include: *Ascyrum multicaule, Aster turbinellus, Brachyeletrum erectum, Bromus purgans, Elephantopus carolinianus, Galium pilosum, Hieracium gronovii, Panicum lindheimeri, Panicum linearifolium, Carex glaucodea, Swertia caroliniensis, Silene stellata, Gillenia stipulata, Dodecatheon meadia,* and *Cynoglossum virginianum.*

Dominant Species Association

Species association on mid-slopes is probably most nearly representative of conditions that exist in the average physiographic setting of southern Illinois. With the exception of white oak, none of the lowland species is an important species association constituent on the mid-slope. Three species of oak (white, red, and black), along with the pignut and shagbark hickory, constitute the major portion of the canopy species on mid-slope. Seven species of oaks and hickories were present on the mid-slopes of the twenty study areas (Table 36). Sugar maple is the only lower slope dominant with many species associations on the mid-slope. Here it is associated mostly with white oak and red oak. White oak leads all tree species in number of associations and is followed by black oak and red oak (Table 36).

36. *Association of species, random pairs of trees on mid-slopes*

NO. KEY AND SPECIES	1	2	3	4	5	6	7	8	9	10	11	12	13
1. Quercus alba	65	44	57	3	5	25	15	2	3	1	5	2	15
2. Quercus rubra	44	9	12	3	2	7	9	3	1	2	3	4	12
3. Quercus velutina	57	12	27	1	3	19	9	1	1	3	2	3	3
4. Quercus muhlenbergii	3	3	1	0	0	0	0	0	0	0	1	0	3
5. Liriodendron tulipifera	5	2	3	0	1	3	0	0	1	1	3	4	3
6. Carya glabra	25	7	19	0	3	11	7	2	1	0	3	3	4
7. Carya ovata	15	9	9	3	0	7	9	3	0	1	4	4	7
8. Carya cordiformis	2	3	1	0	0	2	3	1	0	0	2	0	1
9. Carya tomentosa	3	1	1	0	1	1	0	0	0	0	0	1	1
10. Carya ovalis	1	2	3	0	1	0	1	0	0	0	0	1	0
11. Fagus grandifolia	5	3	2	1	3	3	4	2	0	0	4	1	2
12. Nyssa sylvatica	2	4	3	0	4	3	4	0	1	1	1	2	2
13. Acer saccharum	15	12	3	3	3	4	7	1	1	0	2	2	4

Data from 20 areas. Species found less than three times with other species omitted.

In some areas of drier or poorer soil the southern red oak (*Quercus falcata*), post oak, and scarlet oak may become variously mingled with other oaks and hickories. Southward from Atwood Ridge over the Thebes Hills occur oak communities on dissected limestone hills where rock chestnut oak (*Quercus prinus*) is an important dominant. Other associated species here are chinquapin (*Quercus muhlenbergii*), shagbark hickory, red, white, and black oaks, pignut hickory, false shagbark hickory (*Carya ovalis*), sugar maple, sour gum, and, formerly, chestnut (*Castanea dentata*).

Black Oak (Quercus velutina)

This fast growing oak reaches maturity in less than a hundred years. It attains a height of over one hundred feet, and specimens of sixty feet may often be found. Black oak does not persist in a dense overstory and thus is often most abundant in cut-over upland forests where the canopy is incomplete (Brinkman, 1957).

Black oak grows best on mid-slopes where it is most commonly associated with white oak, red oak, southern red oak, and bitternut and pignut hickory. At Jackson Hollow, black oak makes up 42 per cent of the composition and is found most abundantly with white oak (Table 37). At Giant City State Park it reaches 66 per cent. At Union County State Forest, black oak shows an adaptability to the habitat conditions of south slopes where it makes up 49 per cent of the composition of the upland forest.

Flowering of this species is in late April or in May. Black oak is monoecious. Staminate flowers are borne on preceding season's growth and pistillate flowers are borne at the base of the current season's leaf stems (Lamb, 1915). Hybridization with other members of the red oak group is thought to be common (Brinkman, 1957). Acorns ripen in October of the second year and fall to the earth before winter. Germination is high, though acorns often are attacked by nut weevils (Korstian, 1927) and are used readily by wildlife so that a major part of the crop is destroyed.

White Oak (Quercus alba)

The existence of white oak trees 150 feet tall, 8 feet in diameter, and 600 years of age emphasizes the durability, length of life, and grandeur of this oak (Minckler, 1957). This species in southern Illinois is found abundantly on mid- and upper slopes of tributary valleys. At

37. *Percentage composition of trees on selected mid-slope situations*

SPECIES	Midland Hills	Fountain Bluff	Jackson Hollow
Quercus alba	34.58	....	....
Liriodendron tulipifera	1.08	4.12	....
Carya glabra	2.36	1.63	20.71
Quercus rubra	19.49	19.59	4.46
Quercus velutina	37.78	9.61	41.57
Quercus falcata	4.67	....	....
Quercus muhlenbergii	....	9.93	....
Quercus alba	....	41.47	22.30
Fraxinus americana	....	3.51	....
Ulmus rubra	....	0.73	....
Acer saccharum	....	1.86	....
Fagus grandifolia	....	0.73	....
Carya tomentosa	....	2.27	....
Cornus florida	....	0.46	....
Acer nigrum	....	4.47	....
Nyssa sylvatica	....	....	6.39
Carya ovata	....	....	4.40

Little Grand Canyon it comprises 47 per cent of the composition of upper ridges, 43 per cent at Dry Hill, and 37 per cent at Fountain Bluff. It comprises 36 and 39 per cent of mid-slope communities at Beartrack Hollow and Pounds Hollow, respectively (Table 32). It occurs as a member of ravine communities at Midland Hills, Hooven Hollow, Pomona, Lusk Creek, and Jackson Hollow.

White oak, having an occurrence on ravine bottoms, mid- and upper slopes, and a variety of slope aspects, enters into many different communities. It is associated with sugar maple, red oak, beech, tulip tree, black oak, shagbark hickory, pignut hickory, and post oak.

Flowering of white oak is at the time leaves appear and, in our area, this is usually in mid-April. Both male and female flowers are borne on the same tree. Cross fertilization between trees is common (Minckler, 1957). The acorns ripen in September and October and fall continuously during these months. Germination is high and usually takes place soon after falling. White oak seedlings survive under low soil moisture in the top foot of soil, probably as the result of early development of a taproot (Minckler, 1957).

Red Oak (Quercus rubra)

Red oak is widespread in the deciduous forest. It grows on variable terrain and soil conditions but favors northerly or easterly slope aspects and well-drained soils (Sander, 1957). From twenty areas scattered across southern Illinois, the red oak predominated in two lowland communities and was a part of five others. In mid-slope areas it predominated twice and was a part of ten other communities. Red oak did not occur as a leading component of upper or ridge tops and occurred there as a component in only two communities. Thus a mid-slope situation is favored (Table 32). Its commonest associates are white oak, black oak, beech, shagbark hickory, tulip tree, and pignut hickory. It is most often associated with two other oaks—the white and the black (Table 36).

Flowering is before or with the appearance of leaves in late or mid-April. Male flowers are borne in catkins in axils of leaves of the preceding season. Female flowers are single or in several-flowered spikes and are produced in axils of the developing leaves. Fruit or acorns mature in two years and ripen in September and October (Sander, 1957). Over-wintering in the forest litter breaks the natural dormancy and germination takes place in spring following their ripening in fall. About half the acorns falling from trees have been found sound (Downs and McQuilkin, 1944), but animals cause a further great reduction in numbers available for germination. Early survival is tied closely with available soil moisture conditions. Red oak is less tolerant than white oak, but is more tolerant than black oak (Sander, 1957).

Pinus echinata-Carya buckleyi/Vaccinium *Community*

The two sites in southern Illinois where yellow pine (*Pinus echinata*) occurs present somewhat different species composition and topography. These communities are not greatly unlike other ridge communities in southern Illinois in species composition, the only major exception being the presence or absence of yellow pine.

The greatest occurrence of yellow pine in southern Illinois is in the Pine Hills, Union County. The pines are found from mid-slope to near the ridge-tops, becoming progressively more abundant near the top. The ground is a thin layer of rocky, reddish chert, underlain by massive limestone. Yellow pine comprises nearly forty per cent of the canopy, with Buckley's hickory and post oak contributing twenty

THE UPLAND SERIES

per cent each (Table 38). Other canopy species include pignut hickory, white oak, black oak, red oak, black jack oak, and shagbark hickory. An abundant shrub layer is dominated by high-bush blueberry (*Vaccinium vacillans*) and farkleberry (*V. arboreum*). Other common shrubs are shining sumac (*Rhus copallina*), aromatic sumac (*R. aromatica*), New Jersey tea, and wild rose (*Rosa setigera*).

38. *Percentage composition of tree species in selected Pinus echinata-Carya buckleyi/Vaccinium communities*

SPECIES	Pine Hills	Piney Creek
Pinus echinata	40.0	35.5
Carya buckleyi	20.0	35.5
Quercus stellata	20.0	12.2
Carya glabra	5.7	5.6
Quercus alba	4.1	....
Quercus rubra	3.5	....
Quercus velutina	3.5	....
Quercus marilandica	1.7	5.6
Carya ovata	1.5	....
Ulmus alata	....	5.6

Union and Randolph counties respectively.

The herbaceous layer in the *Pinus echinata-Carya buckleyi/Vaccinium* community of the Pine hills is composed predominantly of summer- and fall-flowering species. Early spring brings into flower plantain-leaved everlasting and four species of sedge (*Carex artitecta, C. emmonsii, C. physorhyncha,* and *C. umbellata*). This period is closely followed by blooming of cleft phlox and bird's-foot violet (*Viola pedata*). From mid-May until October, there is a rapid succession of goat's-rue (*Tephrosia virginiana*), bedstraw (*Galium circaezans*), shooting star (*Dodecatheon meadia*), milkweed (*Asclepias quadrifolia*), bee balm (*Monarda bradburiana*), poverty oat-grass (*Danthonia spicata*), pencil flower (*Stylosanthes biflora*), three-seeded mercury (*Acalypha gracilens*), and several species of sunflower and goldenrod. The latter are interesting, floristically, since two species (*Solidago strigosa* and *S. bootii*) are known from no other station in Illinois.

A second area for occurrence of yellow pine is in southeastern Randolph County near West Point. The topography is rolling and the area is drained by Piney Branch of Mill Creek which has cut deeply through sandstone. This resultant wear has created steep-walled canyons and the stream itself has many curves. Though less rocky than those of the Pine Hills, the soils here are gravelly, droughty in summer, reddish in color, and acidic.

Pines occur from the mid-slopes to ridge-top, comprising nearly 35 per cent of the canopy. Buckley's hickory is equally abundant, while post oak, black jack oak, winged elm, and pignut hickory are not infrequent. Some unusual associations of trees were seen to occur on this rough terrain. Beech (*Fagus grandifolia* var. *caroliniana* forma *mollis*) was seen not more than 35 feet from black jack oak and about 10 feet from yellow pine.

A shrub flora in which Carolina buckthorn (*Rhamnus caroliniana*) and blackhaw (*Viburnum rufidulum*) occur is otherwise similar to the mid-layer of the pine stands in Union County. The herbaceous layer has many species which are common to the Appalachian and Salem plateaus. Over half of the characteristic herbaceous species listed by Steyermark (1951) for the thin acid soils underlain by sandstone in Missouri are found here. Thus, this area is strongly related floristically to the Ozarks. Species included in this comparison from this area are: *Pinus echinata, Agrostis elliottiana, Andropogon virginicus, A. scoparius, Festuca octoflora, Panicum sphaerocarpon, Luzula bulbosa, Hypoxis hirsuta, Carya buckleyi, Ranunculus harveyi, Tephrosia virginiana, Danthonia spicata, Ascyrum multicaule, Viola pedata, Pycnanthemum flexuosum, Cassia fasciculata, Stylosanthes biflora,* and *Psoralea psoralioides.*

Quercus stellata-Q. marilantica/Danthonia *Community*

Upon ascending the slopes to near the crest of the hills one is soon aware of the changing microclimatic conditions. The copious shade, attending high humidity, and soothing coolness of the lower slopes is replaced by an environment of greater light, searing summer heat, and lowered humidity. Here on the ridges or crests of the worn sandstone hills where soil is thin, dry, and acidic may be found our most starkly expressed community. Often only two species of trees and seldom more than four or five make up as much as seventy per cent of the composition (Table 39). Locally red cedar may be present and often the winged elm as well. The mid-layer is composed mostly of farkleberry and to a lesser extent the highbush blueberry.

The thin litter crackles underfoot. A pungent odor of red cedar and the dry, dusty smells of the scrub oaks pervade the air. Poverty

THE UPLAND SERIES

oat grass and broomsedge are usually a conspicuous cover, though a thin covering of herbaceous plants is the rule. In rocky crevices may be seen the rock cress (*Sedum pulchellum*), agave, goat's rue, *Houstonia nigricans*, pencil flower, and others. In places, particularly under red cedars of pole size, the reindeer lichen (*Cladonia* sp.) is present as enlarged patches several feet in diameter.

39. *Percentage composition of trees in selected ridge-top tree communities*

SPECIES	Belle Smith Springs	Devil's Kitchen	Hooven Hollow
Carya glabra	4.80	22.10	5.14
Quercus rubra	7.16	....	....
Quercus velutina	8.97	13.22	1.72
Quercus alba	31.73	5.11	6.19
Quercus stellata	25.19	45.23	63.69
Quercus marilandica	9.13	9.80	15.80
Juniperus virginiana	11.52	....	2.13
Ulmus alata	1.29	....	1.94
Carya ovata	....	4.52	0.61
Carya buckleyi	....	....	2.81

Pope, Williamson, and Hardin Counties respectively.

On 20 upland areas post oak averaged 31.4 per cent composition and had a frequency of 44.5 per cent. White oak averaged 18.8 per cent of the upland composition and presented a frequency of 28.1 per cent for the twenty areas. Black oak averaged 12.4 per cent composition and 18.2 per cent frequency (Table 40).

Dominant Species Association

White oak is much less abundant on the ridge-top although it still ranks second in number of association combinations. Post oak holds an undisputed first place on the ridge-top (Table 41). It is not only the most frequent associate with itself, but is likewise the most frequent associate with the majority of the other upland species. The figures show that pignut is the leading hickory.

Post Oak (Quercus stellata)

Post oak, a frequent companion of the black jack, is a tree of slightly greater size. These two tree species contrast in bark color as well as in size. Post oak, because of its gray bark, is reminiscent of white oak. Usually broad but lighter colored patches of bark appear at intervals on older trees. Leaves are usually 4 to 6 inches long, 3 to 4

40. *Percentage of frequency and composition and number of areas in which selected species were present on uplands*

SPECIES	Frequency	Composition	Areas
Quercus stellata	44.45	31.35	18
Quercus alba	28.11	18.84	19
Quercus velutina	18.22	12.38	17
Carya glabra	20.83	10.59	20
Quercus marilandica	12.22	6.43	12
Juniperus virginiana	12.33	5.66	8
Quercus rubra	0.17	4.12	1
Carya ovata	7.49	3.67	18
Ulmus alata	4.11	1.22	10
Fraxinus americana	2.33	1.19	7
Carya ovalis	1.17	0.87	2
Acer saccharum	1.50	0.69	4
Ulmus americana	0.67	0.51	2
Quercus muhlenbergii	0.58	0.42	3
Juglans nigra	0.80	0.36	5
Carya buckleyi	0.64	0.27	3
Nyssa sylvatica	0.33	0.20	2
Carya tomentosa	0.67	0.18	2
Pinus echinata	0.13	0.12	1
Acer rubrum	0.13	0.07	1
Ostrya virginiana	0.33	0.06	1
Prunus serotina	0.17	0.06	1
Cornus spp.	0.33	0.03	2
Fagus grandifolia	0.17	0.03	1
Acer saccharum	0.17	0.02	1
Morus rubra	0.17	0.02	1

Data from 20 areas.

inches wide, broadest above the middle, and deeply lobed into 5 lobes. The upper 3 lobes are most conspicuous giving the upper part of the leaf an appearance of a cross. Both staminate and pistillate flowers are borne separately on the same tree. Staminate catkins are 3 to 4 inches long appearing with the leaves in May. Pistillate flowers are single on short stalks. Acorns are included for about ⅓ their length into cups and the overall length is from ½ to 1 inch.

Post oak on better sites will be companion to black oak, white oak, and hickory. It will also be found on wet soil. Thus its usual

THE UPLAND SERIES

associates in southern Illinois are black oak, black jack oak, hickory, pin oak, elm, and ash. Post oak and black jack oak have long been used as indicators of poor land. Indeed these trees were often used in earlier days for fixing tax rates on land. Post oaks because of their slow growth yield little lumber. When cut for use as mine props and fence posts they barely pay taxes on the land (Miller and Tehon, 1929).

Black Jack Oak (Quercus marilandica)

Black jack oak is a small tree of 20 or 30 feet in height. It has an open, rounded crown, a blackish bark, and inhabits sandy or clayey barren lands. It is known also as scrub oak. Leaves are about 6 inches long and sometimes quite as broad. They are leathery, dark green, and broadest at about ⅔ the distance from where the blade joins the petiole. At its broadest point the leaf is characterized by 3 lobes which are not pronounced. The leaves taper from the lateral lobes to a rounded base.

Sexes are separate. On a single tree the staminate flowers are borne in catkins of 2 to 4 inches length which appear in May with the leaves. The pistillate flowers are often in pairs appearing at the ends of rusty-colored pubescent pedicels. Up to ⅔ of the acorn is included within the cup which is stalked. The overall length of the acorn is about ¾ of an inch.

1. *Association of species, random pairs of trees on ridge tops*

KEY AND SPECIES	1	2	3	4	5	6	7	8	9	10	11	12	13
Quercus stellata	82	39	23	19	5	38	2	0	17	0	27	7	5
Quercus marilandica	39	14	10	4	1	5	1	0	0	0	3	0	0
Quercus alba	23	10	40	34	19	25	0	0	4	4	5	0	0
Quercus velutina	19	4	34	12	6	20	0	0	5	1	1	1	3
Quercus rubra	5	1	19	6	5	2	1	0	4	2	0	2	0
Carya glabra	38	5	25	20	2	15	0	1	10	0	8	2	0
Carya buckleyi	2	1	0	0	1	0	0	0	0	0	0	0	0
Carya tomentosa	0	0	0	0	0	1	0	1	0	0	0	1	0
Carya ovata	17	0	4	5	4	10	0	0	3	0	1	0	3
Carya ovalis	0	0	4	1	2	0	0	0	0	0	0	6	0
Juniperus virginiana	27	3	5	1	0	8	0	0	1	0	10	6	0
Ulmus alata	7	0	0	1	2	2	0	1	0	6	6	0	0
Fraxinus americana	5	0	0	3	0	0	0	0	3	0	0	2	0

Data from 20 areas. Species found less than three times with other species omitted.

Tree Reproduction on Uplands

Eighteen hundred meter-wide transects between randomly selected trees in twenty upland study areas were taken in which the number of individuals of various shrubs and trees was recorded by four diameter size classes. These size classes were (1) 0 to 0.9 inch; (2) 1.0 to 1.9 inches; (3) 2.0 to 2.9 inches; (4) 3.0 to 3.9 inches.

Examination of these data (Table 42) reveals that the leading canopy species previously discussed are not always leaders in tree reproduction size classes. This would emphasize that these forests are still developing and that further changes in the succeeding generations are still to occur before the stability of a climax is reached. A few species, most frequently in smaller size classes, have all but disappeared in the larger size classes. Competition and reaction would readily account for most of this; however, some of the most shade tolerant species are the ones to show the lowest rate of reproduction; thus, factors other than shade are operant.

In the small size class (0 to 0.9 inch), sugar maple far outnumbers all other species on lowland and mid-slopes. Among canopy species, American elm and hickories are next and the mid-layer components of blue beech, spicebush, and bladdernut are abundant. On mid-slopes, American elm, sugar maple, hickories, and oaks become most important in order. On upland ridges it is winged elm and the hickories which show the greatest numbers in small size reproduction and these are followed by post oak and white oak.

On lowland and mid-slopes in the second size class of reproduction (1.0 to 1.9 inches), sugar maple still ranks highest by a considerable margin. Hickories and oaks rank next in order. Winged elm leads all other individuals on upland in this size class followed closely by hickory.

In the third reproduction size class group (2.0 to 2.9 inches), there is a continued reduction in number among all species. Sugar maple remains a leading species on lowland with hickories and oaks leading in number on upland. Reduction is in number of the relatively intolerant trees such as winged elm and white ash and also the tolerant sugar maple.

In the large reproduction size class (3.0 to 3.9 inches), sugar maple leads on lowland and hickories and oaks are most abundant on mid-slopes and upland ridges. Beech reproduction in all size classes is

42. *Shrub and tree reproduction in selected situations, diameter in inches at one foot above ground*

SPECIES	Lowlands				Mid-slopes				Upland Ridges			
	0–.9	1–1.9	2–2.9	3–3.9	0–.9	1–1.9	2–2.9	3–3.9	0–.9	1–1.9	2–2.9	3–3.9
Acer negundo	1	..	..	..	..	..	..	..	..	..	..	..
Acer rubrum	18	3	..	..	2	3	..	..	10	3	..	..
Acer saccharum	97	133	32	13	73	118	14	7	13	12	..	1
Amelanchier arborea	3	3	..	..	3	3	..	..	4	6	1	..
Aralia spinosa	1	..	..	..	1	..	..	..	..	..	..	..
Asimina triloba	31	8	..	..	10	4	1	..	2	..	..	..
Carpinus caroliniana	7	9	8	1	1	2	..	..	..	..	..	..
Carya spp.	46	25	12	1	51	72	24	10	71	49	20	7
Celtis occidentalis	23	3	5	1	30	15	2	1	3	7	..	..
Cornus florida	22	15	5	2	20	27	5	7	13	22	6	1
Corylus americana	3	..	..	..	2	1	..	..	..	..	..	..
Crataegus spp.	..	..	..	..	1	..	..	..	2	1	..	..
Diospyros virginiana	2	..	..	..	6	1	..	..	3	..	1	..
Fagus grandifolia	12	5	3	1	..	3	..	1	..	..	..	..
Fraxinus americana	27	10	3	..	19	25	..	..	7	8	3	..
Juniperus virginiana	7	..	..	..	7	3	..	..	25	16	3	3
Juglans nigra	..	..	..	..	1	..	..	..	..	..	..	..
Juglans cinerea	..	..	..	..	2	..	..	..	..	..	..	..
Lindera benzoin	47	..	..	..	6	..	..	..	..	..	..	..
Liriodendron tulipifera	16	1	..	..	4	1	..	1	..	..	..	..
Liquidambar styraciflua	1	..	..	..	..	..	..	..	..	..	..	..
Morus rubra	9	2	1	..	11	7	..	..	1	1	..	..

SPECIES	Lowlands				Mid-slopes				Upland Ridges			
	0–.9	1–1.9	2–2.9	3–3.9	0–.9	1–1.9	2–2.9	3–3.9	0–.9	1–1.9	2–2.9	3–3.9
Nyssa sylvatica	30	10	3	2	25	6	2	1	1	1	..	..
Ostrya virginiana	67	38	14	6	52	30	8	3	20	24	5	1
Prunus serotina	8	..	..	..	12	1	..	..	4	..	..	..
Prunus lanata	..	..	..	..	1	..	..	..	..	..	..	..
Pinus echinata	..	..	..	..	1	..	..	..	..	..	..	..
Quercus alba	20	5	5	2	38	20	15	4	25	14	5	2
Quercus muhlenbergii	4	5	11	..	3	2	1	1	1	..	..	..
Quercus rubra	19	1	..	1	37	4	1	..	11	3	..	..
Quercus marilandica	..	..	..	..	1	..	..	..	14	4	3	..
Quercus stellata	..	..	..	..	6	1	..	..	25	11	11	2
Quercus velutina	..	..	..	..	17	5	2	..	8	8	1	1
Sassafras albidum	45	2	..	..	57	12	1	..	25	10	..	..
Staphylea trifolia	47	4	..	..	..	..	..	..	..	..	..	..
Ulmus alata	1	1	..	..	..	6	1	..	61	53	14	4
Ulmus americana	88	10	5	1	91	9	1	3	9	..	..	..
Ulmus rubra	8	1	..	..	5	4	..	..	..	..	..	..
Vaccinium arboreum	12	1	1	..	4	3	..	..	14	31	1	1
Viburnum prunifolium	..	..	..	..	..	1	..	..	1	..	..	..
Tilia americana	..	..	..	..	..	..	..	..	1	..	..	..

Total individuals in 1,800 meter-wide transects between randomly selected trees.

surprisingly low. Dogwood and blue beech, two of the mid-layer species, are well represented in all size classes. These data suggest a presently developing vegetation for southern Illinois area uplands of oaks and hickories.

Annotated List of Upland Forest Areas

Fountain Bluff	Sects. 1, 6, 7, 12, 13, 18, T 10 S, R 3 W	Jackson Co.
	Sects. 31, 36, T 9 S, R 3 W	Jackson Co.
Dry Hill	Sect. 21, T 8 S, R 4 W	Jackson Co.
Midland Hills	Sect. 18, T 10 S, R 1 W	Jackson Co.
Piney Creek	Sect. 23, T 7 S, R 5 W	Randolph Co.
Pomona Natural Bridge	Sect. 17, T 10 S, R 2 W	Jackson Co.
Little Grand Canyon	Sect. 1, T 10 S, R 3 W	Jackson Co.
Giant City	Sects. 27, 28, 33, 34, T 10 S, R 1 W	Jackson Co. and Union Co.
Pine Hills	Sect. 16, T 11 S, R 2 W	Union Co.
	Sect. 9, T 11 S, R 3 S	Union Co.
Panthers Den	Sect. 35, T 11 S, R 1 E	Union Co.
Ferne Clyffe	Sects. 22, 27, T 11 S, R 2 E	Johnson Co.
Devils Kitchen	Sects. 16, 21, T 10 S, R 1 E	Williamson Co.
Belle Smith Springs	Sect. 33, T 11 S, R 5 E	Pope Co.
Dixon Springs	Sects. 28, 29, T 13 S, R 5 W	Pope Co.
Beartrack Hollow	Sects. 7, 17, 18, T 11 S, R 6 E	Pope Co.
Grindstaff Hollow	Sects. 21, 22, 27, 28, T 10 S, R 8 E	Pope Co.
Jackson Hollow	Sect. 18, T 11 S, R 5 E	Pope Co.
Pounds Hollow	Sects. 25, 35, 36, T 10 S, R 8 E	Hardin Co.
Hooven Hollow	Sect. 22, T 11 S, R 9 E	Hardin Co.
Blind Hollow	Sect. 18, T 11 S, R 10 E	Hardin Co.

LITERATURE CITED

Anonymous. 1921. Huge lone sycamore discovered. Amer. For. 27:698.
———. 1959. Yellow-poplar seed quality varies by seed trees, stands, and years. Central States For. Expt. Stat., Columbus, Ohio. note 134.
Blake, A. K. 1935. Viability and germination of seeds and early life history of prairie plants. Ecol. Monog. 5:405–60.
Bollwinkel, Carl W. 1954. Sink-hole pond vegetation of southern Illinois. Master's thesis, Southern Illinois University (unpublished).
Bonnell, Clarence. 1946. The Illinois Ozarks. Harrisburg, Ill.: publ. by the author.
Braun, E. Lucy. 1943. An annotated catalog of Spermatophytes of Kentucky. Publ. by the author.
———. 1950. Deciduous forests of eastern North America. Philadelphia: The Blakiston Co.
Brendel, F. 1887. Flora Peoriana. Peoria, Ill.
Brinkman, Kenneth A. 1957. Silvical characteristics of black oak. Central States For. Expt. Stat., U.S.D.A. For. Service Misc. Release 19.
Britton, N. L., and A. L. Brown. 1923. The New Illustrated Flora. 2nd ed. Lancaster, Pennsylvania.
Buell, M. 1949. Gründiss der Krauterkünde. Ecol. 20 (1).
Cain, S. A. 1930. Certain floristic affinities of the trees and shrubs of the Great Smoky Mountains and vicinity. Butler Univ. Bot. Stud. 1 (9):129–56.
———. 1944. Foundations of plant geography. New York: Harper & Brothers.
Canfield, R. 1941. Application of the line interception method in sampling range vegetation. Jour. For. 39:388–94.
Clements, F. E. 1904. Research methods in ecology. Lincoln, Neb.: Univ. Publ. Co.
———. 1916. Plant succession. Carnegie Inst. Wash. Publ. 242.
Clements, F. E., and V. E. Shelford. 1939. Bio-ecology. New York: John Wiley and Sons.

Climate and Man. 1941. Yearbook of agriculture. U.S.D.A., Washington, D. C.: Gov't. Printing Office.

Conard, Henry S. 1951. The background of plant ecology. Iowa State College Press.

Cottam, G., and J. T. Curtis. 1949. A method for making rapid surveys of woodlands by means of pairs of randomly selected trees. Ecol. 30: 101–4.

―――. 1955. Correction for various exclusion angles in the random pairs method. Ecol. 36:767.

―――. 1956. The use of distance measures in phytosociological sampling. Ecol. 37:451–60.

Coupland, R. T. 1950. Ecology of mixed prairie in Canada. Ecol. Monog. 30:271–315.

Cowles, H. C. 1899. The ecological relations of the vegetation on the sand dunes of Lake Michigan. Bot. Gaz. 27:95–116, 167–202, 281–308, 361–91.

―――. 1901. The plant societies of Chicago and vicinity. Geog. Soc. Chicago Bull. 2:13–76.

―――. 1901a. The physiographic ecology of Chicago and vicinity: A study of the origins, development, and classification of plant societies. Bot. Gaz. 31:73–108, 145–82.

―――. 1929. The succession point of view in floristics. Proc. Int. Congress Plant Sci., Ithaca, N. Y. I:687–91.

Curtis, J. T., and R. P. McIntosh. 1951. An upland forest continuum in the prairie-forest border region of Wisconsin. Ecol. 32:476–96.

Daubenmire, R. 1952. Forest vegetation of northern Idaho and adjacent Washington and its bearing on concepts of vegetation classification. Ecol. Monog. 22:301–30.

Deam, C. C. 1940. Flora of Indiana. Indianapolis: Dept. Cons., Div. Forestry.

Demaree, D. 1932. Submerging experiments with Taxodium. Ecol. 13: 258–62.

Desmarais, Yves. 1942. Dynamics of leaf variation in the sugar maples. Brittonia 7:347–88.

Downs, A. A., and W. E. McQuilkin. 1944. Seed production of southern Appalachian oaks. Jour. For. 42:913–20.

Eddy, S. 1931. The plankton of some sink-hole ponds in southern Illinois. Bull. Ill. Nat. Hist. Surv. 19:449–59.

Eifert, V. 1959. River World. New York: Dodd, Mead, and Co.

Engelmann, A. A. 1866. Geological survey of Illinois. Vol. I.

―――. 1868. Geological survey of Illinois. Vol. III.

Evers, R. A. 1955. Hill prairies of Illinois. Ill. Nat. Hist. Surv. Bull. 26: 367–446.

―――. 1961. The Filmy fern in Illinois. Ill. Nat. Hist. Surv. Biol. notes. No. 44.

LITERATURE CITED

Fenneman, N. M. 1938. Physiography of eastern United States. New York: McGraw-Hill.

Fritts, H. C., and B. J. Kirkland. 1960. The distribution of river birch in Cumberland County, Illinois. Trans. Ill. Acad. Sci. 2:68–70.

Gannett, Henry. 1892. The average elevation of the United States. U.S. Geol. Surv. 13th Ann. Rpt.

Garner, W. W., and A. A. Allard. 1920. Effect of the relative length of day and night and other factors of the environment on growth and reproduction in plants. Jour. Agr. Res. 18:553–606.

Gleason, H. A. 1923. The vegetational history of the Middle West. Ann. Assoc. Amer. Geogr. 12:39–85.

———. 1926. The individualistic concept of the plant association. Bull. Torrey Club 53:71–126.

Goldthwait, R. P. 1959. Scenes in Ohio during the last ice age. Ohio Jour. Sci. 59 (4):193.

Grosenbaugh, L. R. 1952. Plotless timbers estimates—new, fast, and easy. Jour. For. 50:32–37.

Gunning, G., and W. Lewis. 1955. The fish population of a spring-fed swamp in the Mississippi bottoms of southern Illinois. Ecol. 36 (4): 552–558.

Harlow, W. M. and E. S. Harrar. 1950. Textbook of Dendrology. 3rd ed. New York.

Hopkins, A. D. 1938. Bioclimatics, a science of life and climate relations. U.S.D.A. Misc. Publ. no. 280.

Horberg, L. 1950. Bedrock topography of Illinois. Ill. Geol. Surv. Bull. 73. Urbana.

Hosner, J. F. 1957. Effects of water upon the seed germination of bottomland trees. Forest Sci. 3 (1):67–70.

Hus, H. 1908. An ecological cross section of the Mississippi River in the region of St. Louis, Mo. Ann. Rept. Mo. Bot. Gard. 19:127–258.

Jones, G. N. 1950. Flora of Illinois. 2nd ed. Notre Dame, Ind.: Univ. Notre Dame Press.

Jones, G. N., and G. D. Fuller, H. E. Ahles, G. S. Winterringer, and A. Flynn. 1955. The vascular plants of Illinois. Museum Sci. Ser. 6. Urbana and Springfield: Univ. of Ill. Press and Ill. State Museum.

Joos, L. A. 1959. Climates of the states: Illinois. U.S. Weather Bureau. Climatography of the United States. No. 60–11.

Korstian, C. F. 1927. Factors controlling germination and early survival in oaks. Yale Univ. School of Forestry Bull. 19.

Lamb. G. N. 1915. A calendar of the leafing, flowering and seeding of the common trees of the eastern United States. U.S. Weather Bur. Monthly Weather Rev. Sup. 2, Part I.

Leighton, M. M., G. E. Ekblaw, and L. Horberg. 1948. Physiographic divisions of Illinois. Jour. Geol. 56:16–33.

Leopold, A. 1949. A sand county almanac. New York: Oxford Univ. Press.

Leopold, A., and S. E. Jones. 1945. A phenological record for Sauk and Dane Counties, Wisconsin. Ecol. Monog. 17:81–122.

Levy, E. B., and E. A. Madden. 1933. The point method of pasture analysis. N.Z. Jour. Agric. 46:267–79.

Logan, W. N. 1922. Handbook of Indiana geology. State Div. of Geol., Indianapolis.

Lutz, H. J. 1930. The vegetation of Heart's Content, a virgin forest in northwestern Pennsylvania. Ecol. 11:1–29.

Merriam, C. H. 1898. Life zones and crop zones of the United States. U.S.D.A., Div. Biol. Surv. Bull. 10.

Merz, R. W. 1958. Silvical characteristics of American sycamore. Central States For. Expt. Sta. Miss. U.S.D.A. Forest Service Release 26.

Miller, R. B., and L. R. Tehon. 1929. The native and naturalized trees of Illinois. Urbana: Dept. of Reg. and Educ., Div. of the Nat. Hist. Surv.

Minckler, L. S. 1957. Silvical characteristics of white oak. Central States For. Expt. Sta. U.S.DA. Forest Service Misc. Release 21.

———. 1957a. Silvical characteristics of pin oak. Central States For. Expt. Sta. U.S.D.A. Forest Service Misc. Release 20.

Mohlenbrock, R. H. 1959. A floristic study of a southern Illinois swampy area. Ohio Jour. Sci. 59:89–100.

———. 1953. Flowering plants and ferns of Giant City State Park. Springfield: Div. of Parks and Memorials, Dept. of Conservation and the Ill. State Museum.

Mohlenbrock, R. H., and J. W. Voigt. 1959. A flora of southern Illinois. Carbondale: So. Ill. Univ. Press.

Nichols, G. E. 1923. A working basis for the ecological classification of plant communities. Ecol. 4:11–23, 154–79.

Oehser, P. H. 1959. The word ecology. Sci. 129, no. 3355.

Oosting, H. J., and L. E. Anderson. 1939. Plant succession on granite rock in eastern North America. Bot. Gaz. 100:750–68.

Palmer, E. J., and J. A. Steyermark. 1935. An annotated catalogue of the flowering plants of Missouri. Ann. Mo. Bot. Gard. 22:375–758.

Parker, Dorothy. 1936. Affinities of the flora of Indiana. I. Amer. Midl. Nat. 17:706–24.

Peattie, D. C. 1922. The coastal plain element in the flora of the Great Lakes. Rhodora 24:57–70, 80–83.

Penfound, W. T. 1952. Southern swamps and marshes. Bot. Rev. 18:413–46.

Phillips, E. A. 1959. Methods of vegetation study. New York: Henry Holt and Co.

Potential water resources of southern Illinois. 1957. Report of investigation 31. Urbana: Ill. State Water Surv.

Potzger, J. E., and R. C. Friesner. 1940. A phytosociological study of the herbaceous plants in two types of forests in central Indiana. Butler Univ. Bot. Studies 4:163–80.

LITERATURE CITED

Pound, R., and F. E. Clements. 1898. Phytogeography of Nebraska. 2nd ed. Lincoln: Jacob North and Co.

Raunkaier, C. 1934. The life forms of plants and statistical plant geography. New York: Oxford University Press.

Rensing, M. 1957. The vegetation of the Shawnee hiking trail. Master's thesis. Southern Illinois Univ. (unpublished).

Ridgway, R. 1872. Notes on the vegetation of the lower Wabash valley. Amer. Nat. 6:658–65.

———. 1872b. Notes on the vegetation of the lower Wabash valley. II. Peculiar features of the bottom lands. Amer. Nat. 6:724–32.

Risely, R. G. 1911. Wabash County biographical, big sycamore tree. Chicago: Munsell Publ. Co.

Rowe, J. S. 1956. Uses of undergrowth plant species in forestry. Ecol. 37 (3):461–73.

Sampson, A. W. 1952. Range management principles and practices. New York: John Wiley and Sons, Inc.

Sander, I. L. 1957. Silvical characteristics of northern red oak. Central States For. Expt. Sta. U.S.D.A. Forest Service Misc. Release 17.

Sargent, C. S. 1933. Manual of the trees of North America. Cambridge, Mass.: Houghton Mifflin Co.

Schimper, A. F. W. 1903. Plant geography upon a physiological basis. Oxford: Clarendon Press.

Sears, P. B. 1949. Integration at the community level. Amer. Scientist 37 (2):235–42.

Shanks, R. E. 1953. Forest composition and species association in the beech-maple forest region of western Ohio. Ecol. 34:455–66.

Shantz, H. L. 1917. Plant succession on abandoned roads in eastern Colorado. Jour. Ecol. 5:19–42.

Sinnott, E. W., and K. S. Wilson. 1955. Botany: Principles and Problems. 5th ed. McGraw-Hill Book Co., Inc.

Smith, N. F. 1953. Swamp white oak. Mich. Conserv. 22 (4):31–32.

Steyermark, J. A. 1951. Botanical areas in the Missouri Ozarks. Mo. Bot. Gard. Bull. 39:126–35.

———. 1955. The Ozarks—their past, present, and future. Mo. Bot. Gard. Bull. 43:2–12.

Tansley, A. G. 1935. The use and abuse of vegetational concepts and terms. Ecol. 16:284–307.

Thomas, D. 1816. Travels through western country in the summer of 1816. Indiana Historical Collection.

Tourney, J. W. 1931. Seeding and planting in the practice of forestry. New York: John Wiley and Sons, Inc.

Visher, S. S. 1944. Climate of Indiana. Indiana Univ. Publ. Series no. 13. Bloomington.

Voigt, J. W., and J. E. Weaver. 1951. Range condition classes of native midwestern pasture: an ecological analysis. Ecol. Monog. 21:390–460.

Weaver, J. E. 1954. North American prairie. Lincoln, Nebr.: Johnsen Publ. Co.

Weaver, J. E., and T. J. Fitzpatrick. 1932. Ecology and relative importance of the dominants of tall-grass prairie. Bot. Gaz. 93 (2):113–50.

———. 1934. The prairie. Ecol. Monog. 4:109–295.

Weaver, J. E., and F. W. Albertson. 1956. Grasslands of the great plains. Lincoln, Nebr.: Johnsen Publ. Co.

Weller, S. 1926. The making of southern Illinois. Trans. Ill. Acad. Sci. 19:27–49.

Went, F. W. 1944. Thermoperiodicity in growth and fruiting of the tomato. Amer. Jour. Bot. 31:135–50.

Whitaker, R. A. 1953. A consideration of climax theory: the climax as a population and pattern. Ecol. Monog. 23:41–78.

Winterringer, G. S., and A. G. Vestal. 1956. Rock-ledge vegetation in southern Illinois. Ecol. Monog. 26:105–30.

Wright, J. W. 1959. Silvical characteristics of green ash. Northeastern For. Expt. Sta. Forest Service U.S.D.A. Paper 126.

Yeager, J. E. 1949. Effect of permanent flooding in a river-bottom timber area. Bull. Ill. Nat. Hist. Surv. 25:33–65.

INDEX

Agave, (*Agave*), 57, 183
Alder, (*Alnus*), 104
Alternation, 64
Alum Root, (*Heuchera*), 33
American Bottoms, 138
Appalachians, 25, 31, 32, 33, 36, 41
Arctotertiary Forest, 31
Arrow-arum, (*Peltandra*), 87
Arrowhead, (*Sagittaria*), 132
Arrow-wood, (*Viburnum*), 132
Ash, (*Carya*): Blue, 145; Green, 95, 96; Pumpkin, 26, 86, 92, 94
Aster, (*Aster*), 88, 100, 112, 115, 121
Atwood Ridge, 178
Auroras Bend, 100
Autumn Color, 19

Balance, Nature, 66
Bald Cypress, (*Taxodium*), 26, 37, 86, 89, 92, 94
Basswood, (*Tilia*), 100
Bay Creek, 6, 88
Beards-tongue, Ozark, (*Penstemon*), 36
Beartrack Hollow, 144, 179, 189
Bedrock: Devonian, 5, 6, 22; Mississippian, 5, 22, 25; Pennsylvanian, 5, 22, 25, 26
Bedstraw, (*Galium*), 35, 130, 181
Bee Balm, (*Monarda*), 181
Beech, (*Fagus*), 123–25, 173–74, 182, 186

Beech, Blue, (*Carpinus*), 186, 189
Beech-drops, (*Epifagus*), 172
Beggar's-tick, (*Desmodium*), 161
Belle Smith Springs, 34, 140, 167, 189
Bent-grass, Creeping, (*Agrostis*), 88
Big Muddy River, 6
Bioclimatic Law, 17
Biotic Pyramid, 66
Birch, River, (*Betula*), 107
Bishop's Cap, (*Mitella*), 37, 126, 144
Bittersweet, (*Celastrus*), 175
Black Cohosh, (*Cimicifuga*), 33
Black-eyed Susan, (*Rudbeckia*), 162

Black Slough, 88
Bladder-fern, bulblet, (*Cystopteris*), 146
Bladdernut, (*Staphylea*), 125, 176, 187
Blazing-star, (*Liatris*), 146, 151, 162, 166
Blind Hollow, 144, 189
Bloodroot, (*Sanguinaria*), 14, 125, 172
Bluebell, (*Mertensia*), 172
Blueberry, Highbush, (*Vaccinium*), 181
Blue-eyed Grass, (*Sisyrinchium*), 151, 160
Blue-eyed Mary, (*Collinsia*), 126
Bluegrass, (*Poa*), 133
Bluestem, (*Andropogon*): Big, 38, 154, 155, 156; Little, 38, 141, 155, 157

INDEX

Bluet, Shrubby, (*Houstonia*), 146
Bluff Springs, 158
Boneset, (*Kuhnia*), 162
Bottomland Forest, 97
Box Elder, (*Acer*), 96, 121
Brookweed, (*Samolus*), 130
Broomsedge, (*Andropogon*), 141, 172, 183
Buckbrush, (*Symphoricarpos*), 57
Buckthorn, Carolina, (*Rhamnus*), 26, 143, 145, 182
Buckthorn, Southern, (*Bumelia*), 37
Buffalo Grass, (*Buchloe*), 149
Bulrush, (*Scirpus*), 55, 130
Bur Cucumber, (*Sicyos*), 145
Burden Falls, 140
Buttercup, (*Ranunculus*); Harvey's, 35; Small, 133; Swamp, 104
Butterfly Weed, (*Asclepias*), 161
Buttonbush, (*Cephalanthus*), 56, 86, 88

Cache River, 6, 7, 88
Cactus, (*Opuntia*), 140
Cane, (*Arundinaria*), 26, 97, 103
Cardinal Flower, (*Lobelia*), 97
Catalpa, (*Catalpa*), 101
Catbriar, (*Smilax*), 57
Cattail, (*Typha*), 55, 132
Cave Creek, 167
Cave Hill, 140
Cedar, Red, (*Juniperus*), 57, 140, 145, 183
Central Lowland, 3, 8
Chestnut, (*Castanea*), 178
Chokeberry, Black, (*Aronia*), 143
Clearweed, (*Pilea*), 144
Cleft Phlox, (*Phlox*), 141, 145
Cliff-brake, Purple, (*Pellaea*), 146
Climate, 8, 9
Climax, 53
Clubmoss, (*Lycopodium*), 144, 146
Coastal Plain, 6, 39, 60, 62, 89, 91, 92, 94, 100
Columbine, (*Aquilegia*), 141
Compass-plant, (*Silphium*), 153

Coneflower, Ozark, (*Rudbeckia*), 36
Coneflower, Purple, (*Echinacea*), 156, 161
Continuum, 73
Coontail, (*Ceratophyllum*), 86
Cottonwood, (*Populus*): Common, 56, 96, 98, 99, 104, 106; Swamp, 26, 97
Crab Apple, Prairie, (*Malus*), 155
Crab Orchard Creek, 6
Cress, Bulbous, (*Cardamine*), 133
Cretaceous, 26, 27, 98
Croton, (*Croton*), 142
Culver Root, (*Veronicastrum*), 152
Cup-plant, (*Silphium*), 152
Cutgrass, (*Leersia*), 88, 93, 112

Dandelion, False, (*Krigia*), 131, 142
Daylength, 17, 71
Deep Swamps, 85
Delphinium, (*Delphinium*), 128
Devil's Backbone, 5
Devil's Bake-oven, 5, 23

Devil's Kitchen, 144, 189
Dittany, (*Cunila*), 141
Dixon Springs, 144, 189
Dogwood, (*Cornus*): Flowering, 27, 175, 189; Gray, 97; Rough-leaved, 155
Dropseed, Tall, (*Sporobolus*), 160
Dry Hill, 144, 180, 189
Duckweed, (*Lemna*), 55, 133
Dutchman's Breeches, (*Dicentra*), 125, 172

Ecosystem, 66
Ecotypic Variation, 13
Elder, Marsh, (*Iva*), 99
Elm, (*Ulmus*): American, 97, 186; Slippery, 97; Winged, 140, 142, 186
Elm, Water, (*Planera*), 26
Endemics, 13
Energy Relations, 64

Farkleberry, (*Vaccinium*), 57, 140, 142, 175, 181
Featherfoil, American, (*Hottonia*), 37, 101

INDEX

Fern, Broad Beech, (*Phegopteris*), 173
Fern, Christmas, (*Polystichum*), 172
Fern, Cinnamon, (*Osmunda*), 144
Fern, Filmy, (*Trichomanes*), 33
Fern, Grape, (*Botrichium*), 173
Fern, Hay-scented, (*Dennstaedtia*), 144
Fern, Maidenhair, (*Adiantum*), 173
Fern, Marginal, (*Dryopteris*), 127
Fern, Marsh, (*Dryopteris*), 132
Fern, Polypody, (*Polypodium*), 144
Fern, Regal, (*Osmunda*), 144
Fern, Sensitive, (*Onoclea*), 173
Fern, Walking, (*Camptosorus*), 144, 146
Fern, Water, (*Azolla*), 55
Fern, Woods, (*Woodsia*), 141
Fern, Woolly-lip, (*Cheilanthes*), 141, 142, 143
Ferne Cliffe State Park, 140, 144, 189
Fescue, Six-weeks, (*Festuca*), 141
Feverfew, (*Parthenium*), 152
Finger Grass, (*Digitaria*), 133
Floodplains, 105
Floristics, 41
Flower-of-an-Hour, (*Talinum*), 34, 35, 140, 141
Fountain Bluff, 5, 144, 167, 189
Foxtail, Yellow, (*Setaria*), 151

Garlic, False, (*Nothoscordum*), 131, 141
Geographical Affinity, 41, 42, 43, 44, 45, 46, 47
Geological Time, 22, 23
Giant City State Park, 12, 18, 33, 36, 139, 143, 145, 178, 189
Ginseng, (*Panax*), 173
Glaciation: Illinoian, 29, 38, 112; Kansan, 5, 29; Nebraskan, 29; Wisconsin, 30, 38, 113, 156; mentioned, 12, 27, 28, 29, 37
Goat's-rue, (*Tephrosia*), 182, 183

Goldenrod, Ozark, (*Solidago*), 37, 181
Gold Hill, 140
Government Rock, 162, 166, 167
Grand Tower, 5
Great Smoky Mountains, 33
Grindstaff Hollow, 127, 145, 189
Ground Pine, (*Lycopodium*), 33, 37
Gum, Sour, (*Nyssa*), 178
Gum, Sweet, (*Liquidambar*), 92, 106, 124

Hackberry, Dwarf, (*Celtis*), 27, 125, 145
Harbinger-of-Spring, (*Erigenia*), 17, 125, 126
Harebell, (*Campanula*), 37
Haw, Black, (*Viburnum*), 103, 182
Hawthorn, (*Crataegus*), 97, 115
Heliotrope, Slender, (*Heliotropium*), 36
Hepatica, (*Hepatica*), 126, 127
Hickory, (*Carya*): Buckley's 181, 182; False Shagbark, 178;

Hickory (*continued*) Pignut, 178, 181, 183; Shagbark, 122, 178, 181; Water, 26, 86
Hicks Dome, 22
Holly, (*Ilex*): Deciduous, 26; Verticillate, 37
Hooven Hollow, 145, 179, 189
Hop Hornbeam, (*Ostrya*), 175
Hornwort, (*Ceratophyllum*), 55
Horseshoe Bluff, 32
Horseshoe Lake, 87
Hyacinth, Wild, (*Camassia*), 151

Ice Age, 5, 28, 29, 31
Indian Grass, (*Sorghastrum*), 154, 155
Indian Kitchen, 36, 141
Indian Pipe, (*Monotropa*), 171
Interior Low Plateau, 3
Iris, Swamp (*Iris*), 37, 38
Ivy, Poison, (*Rhus*), 103, 110, 115, 121, 172

Jack-in-the-pulpit, (*Arisaema*), 172
Jackson Hollow, 140, 145, 178, 179, 189

INDEX

June Grass, (*Koeleria*), 149, 155, 160, 166

Land Antiquity, 21
La Rue, 6, 129
Layering, 63
Leek, Wild, (*Allium*), 33, 126
Lespedeza, (*Lespedeza*), 161
Lettuce, Wild, (*Lactuca*), 150
Lichen, 57, 140, 142
Lichen, Reindeer, (*Cladonia*), 183
Life Forms, 69
Lily, Spider, (*Hymenocallis*), 37, 38
Lily, Trout, (*Erythronium*), 126
Line Interception, 76
Lip-fern, Tiny, (*Cheilanthes*), 146
Little Grand Canyon, 126, 145, 189
Lizard's-tail, (*Saururus*), 97, 114
Locust, (*Gleditsia*): Honey, 97; Water, 26, 86, 92
Lost Creek, 134
Lotus, Water, (*Nelumbo*), 55, 86
Lowland Grasses, 146
Lowland Series, 85, 87

Lusk Creek, 141, 179

Magnolia, Cucumber, (*Magnolia*), 35
Mallow, Swamp, (*Hibiscus*), 87
Maple, (*Acer*): Red, 92, 97, 106; Silver, 97, 113, 121; Sugar, 125, 168, 173, 175, 177, 186; Swamp Red, 26
Methods, Vegetation Study, 75
Midland Hills, 145, 179, 189
Milk-Pea, (*Galactia*), 161
Milkweed, (*Acerates*): Green, 161; Woolly, 146
Milkweed, (*Asclepias*): Mead's, 143; Swamp, 97; Whorled, 166
Mimosa, Illinois, (*Desmanthus*), 99
Mississippi Border, 94
Mississippi River, 3, 5, 7, 34, 36, 39, 53, 62, 64, 156
Mistletoe, (*Phoradendron*), 103
Mixed Mesophytic Forest, 62
Moist Rock Walls, 144
Monkey Flower, (*Mimulus*), 97

Moss, 57, 140, 142
Mountain Mint, (*Pycnanthemum*), 166
Muhly, Plains, (*Muhlenbergia*), 158
Mulberry, Red, (*Morus*), 103
Munro-grass, (*Panicum*), 88
Murphysboro, (Lake), 6, 94

Naiad, (*Naias*), 86
Nannyberry, (*Viburnum*) 37
Nature's Design, 49
Needlegrass, (*Stipa*), 149
Nettle, Hedge, (*Stachys*), 34
New Jersey Tea, (*Ceanothus*), 175

Oak, (*Quercus*): Black, 168, 175, 177, 178, 179, 181, 182, 185; Black Jack, 57, 168, 182, 185; Cherrybark, 101, 117; Chestnut, 33, 178; Chinquapin, 178; Overcup, 26, 100, 102; Pin, 106, 114, 115, 116, 117, 122; Post, 57, 168, 178, 183, 184; Red, 168, 175, 177, 178, 179, 181;

Oak (*continued*) Shumard, 100; Southern Red, 178; Spanish, 26; White, 168, 171, 173, 175, 177, 178, 179, 181, 182, 183, 184
Ohio River, 3, 6, 7, 39
Old Stone Face, 163, 164, 167
Olive Branch, 87
Onion (*Allium*), 131
Orchid, Puttyroot (*Aplectrum*), 128
Orchid, Yellow Ladies'-slipper (*Cypripedium*), 128
Ozarks: Arkansas, 35; Illinois, 10, 26, 31, 33, 39, 41; Missouri, 35, 36, 40, 155, 183

Panther's Den, 139, 189
Partridge-berry, (*Mitchella*), 33, 144
Partridge Pea, (*Cassia*), 161, 166
Pawpaw, (*Asimina*), 26, 102
Pecan (*Carya*), 100
Pellitory, (*Parietaria*), 144
Pencil-flower, (*Stylosanthes*), 141, 142, 182

INDEX

Petunia, Wild, (*Ruellia*), 166
Persimmon, (*Diospyros*), 27
Phenology, 13
Phlox, Cleft, (*Phlox*), 14, 145
Pine, (*Pinus*): Shortleaf, 35; Yellow, 180, 181
Pine Hills, 6, 129, 139, 166, 167, 175, 180, 181, 189
Piney Creek, 34, 145, 181, 189
Pinweed, (*Lechea*), 141
Pinxter Flower, (*Rhododendron*), 33
Pipevine, Dutchman's, (*Aristolochia*), 145
Planer Tree, (*Planera*), 86, 92
Plantain, Cordate, (*Plantago*), 104
Plant Communities, 69, 82
Plant Distribution, 11
Pleistocene, 5, 28, 29, 31
Point Quadrat, 78
Pomona, 139, 145, 179, 189
Pond Lily, Yellow, (*Nuphar*), 55, 87
Pondweed, (*Potamogeton*), 86
Pounds Hollow, 145, 179, 189
Poverty-oat grass, (*Danthonia*), 140, 181
Prairie, 6, 27, 38, 54, 71, 146–167
Prairie Clover, (*Petalostemum*): Purple, 38, 161; White, 38, 166
Prairie Dock, (*Silphium*), 152
Prairie Du Rocher, 146, 159, 162, 167
Precipitation, 10
Primrose, Missouri, (*Oenothera*), 35
Privet, Swamp, (*Forestiera*), 97, 102
Puccoon, (*Lithospermum*), 38, 151
Purple-top, (*Tridens*), 150
Pussy-toes, (*Antennaria*), 141

Quadrat, 76
Quillwort, Butler's, (*Isoetes*), 35

Random Pairs, 80
Rattlesnake-master, (*Eryngium*), 151
Ravines, 125
Redbud, (*Cercis*), 26, 175
Rhododendron, (*Rhododendron*), 175
River Bluff, 145
Rock Castle Creek, 34
Rock Cress, (*Sedum*), 140, 142, 183
Rock Ledges, 139
Rock Selaginella, (*Selaginella*), 140
Rock Shelters, 145
Rose, Swamp, (*Rosa*), 88, 94, 95
Rosin-weed, (*Silphium*), 152, 153
Rye, Wild, (*Elymus*), 154

Salem Plateau, 5
Saline River, 6, 129
Saltpeter Cave, 139
Sassafras, (*Sassafras*), 27, 173
Saxifrage, (*Saxifraga*), 143, 144
Seasons, 17, 63, 64
Sedges (*Carex*), 36, 37, 56, 87, 93, 97, 98, 101, 110, 116, 117, 132, 181
Sedum, (*Sedum*), 57
Seepages, 131
Sere: Convergence, 58; Hydrosere, 53, 54, 55; Xerosere, 55, 57; mentioned, 52
Shadscale, (*Amelanchier*), 129
Shawnee Hills, 3, 4, 31, 32
Shawneetown Ridge, 4, 11, 12, 26, 29, 140
Shooting-star, (*Dodecatheon*): Mead's, 181; French's, 13, 126, 128, 144
Side-oats Grama, (*Bouteloua*), 38, 155, 156, 157, 159
Silver Bell, (*Halesia*), 35
Sink-hole Ponds, 134
Site Condition, 105
Slough Grass, (*Spartina*), 152
Smartweed, (*Polygonum*), 97, 98
Snailseed, Carolina, (*Cocculus*), 103, 106, 145
Species Association, 173
Speedwell, (*Veronica*), 133
Spicebush, (*Lindera*), 125, 186
Spiderwort, (*Tradescantia*), 151, 161
Spike-rush, (*Eleocharis*), 56, 133
Spleenwort, (*Asplenium*): Black, 146; Bradley's, 33; Maidenhair, 126
Spleenwort, Narrow-Leaved, (*Athyrium*), 173

Sponge Plant, (*Limnobium*), 37
Spring Beauty, (*Claytonia*), 14, 172
Springs, 127–31
Spurge, Flat-topped, (*Euphorbia*), 166
St. Andrews-cross, (*Ascyrum*), 141
St. John's-wort, (*Hypericum*), 57
Starwort, Water, (*Callitriche*), 130, 132
Stonecrop, (*Sedum*), 34
Stone Mint, (*Cunila*), 176
Storax, (*Styrax*), 37
Storms, 9
Squirrel Corn, (*Dicentra*), 125, 172
Succession, 52, 56, 57, 59, 60, 61
Sumac, (*Rhus*): Aromatic, 175; Shining, 181; Smooth, 172; Staghorn, 137
Swallow Rock, 6
Sweet Flag, (*Acorus*), 87
Switchgrass, (*Panicum*), 38, 153
Sycamore, (*Platanus*), 27, 56, 102, 106, 119

Tear Thumb, (*Polygonum*), 88, 130, 132
Tertiary, 26, 27, 31
Thebes Hills, 178
Thermoperiodicity, 11
Three-seeded Mercury, (*Acalypha*), 176
Tick Clover, (*Desmodium*), 176
Till Plain, 3
Tornado, 10
Trillium, (*Trillium*), 125
Trout Lily, (*Erythronium*), 172
Trumpet Creeper, (*Campsis*), 116
Tulip Tree, (*Liriodendron*), 34, 174
Tupelo, (*Nyssa*), 26, 37, 86, 91, 94
Turtlehead, (*Chelone*), 132

Understory, 105
Union County Forest, 178
Upland Forests, 168

Valerian, (*Valeriana*), 127

Vegetation: Classification, 68; History, 31, 41; Moisture Table, 106; Regions, 40; Structure, 61; mentioned, 48, 49
Vervain, Wild, (*Verbena*), 161
Vines, 103
Violet, (*Viola*): Bird's-Foot, 182; Falcate, 176; Missouri, 133

Wabash Lowland, 7, 96, 98, 99, 101, 108, 109, 111, 119
Wabash River, 3, 7, 39, 119
Wake Robin, (*Trillium*), 172
Walker's Hill, 5
Walker's Pond, 138
Walnut, (*Juglans*), 27
Watercress, (*Nasturtium*), 129, 130
Water Dock, (*Rumex*), 88
Waterleaf, (*Hydrophyllum*), 34, 144
Water Lily, White, (*Nymphaea*), 55

Water Pepper, (*Polygonum*), 88
Water-shield, Carolina, (*Brasenia*), 55, 87
Waterweed, (*Elodea*), 55, 130
Web of Life, 66
Wet Ditches, 132
Whitlow-grass, (*Draba*), 160
Wildcat Hills, 140
Wild Petunia, (*Ruellia*), 156
Williams Hill, 4
Willow, Black, (*Salix*), 56, 96, 97, 98, 104, 106
Willow, Virginia, (*Itea*), 37, 92, 94, 97
Willow, Water, (*Decodon*), 33, 94
Windflower, (*Anemone*), 172
Wing Bluff, 139
Wiregrass, (*Aristida*), 57
Wire Wool, (*Scirpus*), 132
Wolf Lake, 94

Yellowwood, (*Cladrastis*), 27

Zonation, 64